ATC入門

―IFR編―

縄田義直著

鳳文書林出版販売㈱

はしがき

　本書は，これから計器飛行方式によって操縦訓練を行おうとする人を対象とした IFR ATC (ATC Communications) の入門書である．計器飛行方式(IFR: Instrument Flight Rules)とは，「国土交通大臣が指定する空港等からの離陸及びこれに引き続く上昇飛行又は国土交通大臣が指定する空港等への着陸及びそのための降下飛行を，国土交通大臣が定める経路又は国土交通大臣が与える指示による経路により，かつ，その他の飛行の方法について国土交通大臣が与える指示に常時従って行う飛行の方式」(航空法第2条より一部筆者編集・抜粋)であるが，簡単にいうと，あらかじめ定められた経路・方式等で，離陸から着陸までを常時管制官の指示に従って運航することである．よって，計器飛行方式においては，管制官からの指示・情報等が，基本的に英語を用いた無線交信で行われている以上，ATC Communications は非常に重要なものとなってくる．

　本書においては，ある空港からある空港へ運航を行う場合を基本に据え，その場面に付随する用例（そこで行われる可能性がある会話例）を取り上げることにより，IFR ATC Communications に習熟することを目的としている．しかしながら，読者が実際に飛行を行う場合に本書の内容と差異がある場合，運航方式が違う場合はそちらを優先して頂きたい．本書の内容・目的はあくまでも，計器飛行方式での訓練をこれから始めようとする人たちへのイントロダクションであることをご容赦願いたい．

　また，上述した通り，本書は IFR ATC Communications を主眼にしている関係上，一定の飛行経験を有する者が対象となることから，文字の読み方・空域説明・管制機関等の前提知識や VFR 運航関連は扱っていない．また，この本を紐解くまでには既に知っているであろうと思われる知識も紙面の関係上割愛してある．その際には航空専門書籍を参照されたい．

　なお，内容の説明，管制用語の用例等に関しては，以下の文献を参考（一部引用）に行っている．

　社団法人日本航空機操縦士協会　『Aeronautical Information Manual JAPAN』
　国土交通省航空局　『航空保安業務処理規程　第5管制業務処理規程』
　国土交通省航空局　『AIP Aeronautical Information Publication JAPAN』
　財団法人航空振興財団　『国際民間航空条約第1付属書〜18付属書』
　国土交通省航空局　『飛行方式設定基準』

　内容に関して，概ね一般的なものを記述したつもりではあるが，万が一，記述及び内容の間違い等があれば，それはすべて筆者の責に帰すべきものである．

　なお，本書の刊行にあたっては，ディーグラフ横川光陽氏には写真提供で，鳳文書林出版販売青木孝氏に作図・写真提供等で多大なるご協力を頂いた．ここに深謝の意を表したい．

縄田義直

本書の利用の仕方

　本書は，航空管制においてパイロットと管制官（等）が行う通信（英語）に関する用例を収録してある．方式・用語等は日本で規定されているものを使用しているため，外国で訓練する人は，別途，当該国の規程を参照されたい．

　本書は 4 つのパートから構成されている．

　PART.1. では，架空の航空機である Nippon Air 710 が羽田空港から伊丹空港まで運航する場合の交信と，それに関連する事項を扱っている．

　PART.2. では，小型訓練機が計器飛行方式で操縦訓練等を行う場合の交信と，その関連事項を扱っている．

　PART.3. では，2 つのフライトを想定し，管制機関（等）との交信を扱っている．

　PART.4. では，PART.1. 〜 PART.2. で扱わなかった事項を取り上げている．

　本書の対象は初学者ということもあり，説明等はすべて最低限にしている．言葉足らずの場合・意味が不明な場合には別途専門書籍で調べて頂きたい．

　PART.1.，PART.2.，PART.4.（一部）は，概ね以下の構成になっている．

■　Words & Phrases --- 各 UNIT が扱う場面で使用される可能性がある基本的な管制用語を取り扱う（本文に頻繁に出てくる用語は省略している場合もある）．

■　Introduction --- 各 UNIT で扱う内容の，概要説明及び注意すべき点を記してある．

■　Basic Example --- PART.1. においては，P.1 に記載する内容の航空機が行う交信の一例，PART.2. においては，P.85 に記載する内容の航空機の交信を扱っている．

■　Phraseology Example --- Basic Example では出てこない，重要表現等を取扱う．

■　POINT --- 覚えておくとよい最低限の知識を簡潔に記してある．

　本書の利用にあたっては，以下の点に注意願いたい．

■　Nippon Air 710 や Nippon Air 3615 等はすべて架空のコールサインであり，数字に特に意味はない．また，通信・用語に焦点を当てているため，航空機の型式・性能・装備等は特に考慮していない．JA 5806，5807 等は航空大学校で使用している機体のコールサインである．

■　大型機 ＝ PART.1. / 小型機 ＝ PART.2. というわけではなく，単なる便宜上の章立てに過ぎない．PART.1. と PART.2. の両方で 1 つであると考えて頂きたい．

■　表現上統一されていない部分（wind 330 degrees at 5 knots ⇔ wind 330 at 5，QNH 3000 ⇔ QNH 3000 inches，1,500 feet ⇔ 1,500，等）がある．

Table of Contents

はしがき
本書の利用の仕方

PART.3. IFR Flight Scenario

PART.4. Miscellaneous Expressions

<本書におけるチャート類について>

　本書におけるチャート類の一部は,

航空路誌：AIP（Aeronautical Information Publication）

からの抜粋です. チャート類及び作図は, あくまでも本書の説明のために使用しているものであるので, 実際の運航には使用しないようご注意下さい.

<本書における交信について>

　本書における交信は, 実際に行われる又は行われたものではなく, あくまでも模擬したものです. 一部, 実在する名称・航空会社・団体等がありますが, 実際とは一切関係ありません.
　また, 空域再編, 経路・計器進入方式・エンルートチャートの改正, 管制方式に係る記載事項・飛行計画経路の変更等, 及び, 管制サービスのあり方の変更等により, 本書の内容が現状と大幅に異なっている, 又は異なってくる場合もあることを補足しておきます.

<本書で使用している略号等>

区分	略号	表記
管制区管制所	Control	ACC
ターミナル管制所	Radar	RDR
ターミナル管制所入域管制席	Approach / Arrival	APP
ターミナル管制所出域管制席	Departure	DEP
ターミナル管制所 TCA 管制席	TCA	TCA
飛行場管制所	Tower	TWR
飛行場管制所地上管制席	Ground	GND
飛行場管制所管制承認伝達席	Delivery	DEL
着陸誘導管制所	GCA	GCA
飛行場対空援助局	Radio	AFIS
広域対空援助局	Information	AEIS
国際対空通信局	なし（Tokyo）	TOKYO
成田空港ランプコントロール	Ramp	RAMP
飛行援助用航空局	Flight Service	FS

　本文中, 管制機関側で CTL と記載している所は, 特に発話者を特定していない場合です. なお, 上記の中でも, 本書では登場しないものもあります.

PART.1.

IFR Communications - 1

　このPART.1. では，Basic Example で東京国際空港（羽田空港）から大阪国際空港（伊丹空港）へ向かう航空機のシミュレーションを取り上げ，Phraseology Example においてその関連事項を扱う．

　Basic Example におけるフライトの主な内容は以下の通りである．

Basic Example

Aircraft Identification	Nippon Air 710
Type of Aircraft	Boeing 777-200
Departure Aerodrome	Haneda airport
Level	FL 220
Route	LAXAS - Y-56 - TOHME - Y-54 - KOHWA - Y-546 - AGPUK - MIRAI - ABENO - IKOMA
Destination Aerodrome	Osaka airport
SID	LAXAS THREE DEP
STAR	IKOMA EAST ARR
Instrument Approach Procedure	ILS RWY 32L APP

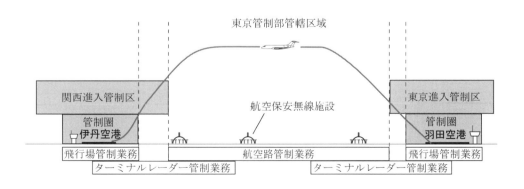

1

UNIT.1. Departures

Words & Phrases

ATC clears / advises / requests	released at ~
管制承認・情報・要求	出発制限を～に解除します
clearance void if not off the ground by ~	
～までに離陸しないときは，この管制承認は無効です	
hold on the ground, expect ~ delay	
地上で待機させて（して）下さい，遅延時間は～の予定です	
EDCT (at) ~ (*1)	revised EDCT (at) ~ (*2)
EDCT は～です	EDCT を～に変更します
EDCT void (*3)	EDCT cancelled (*4)
EDCT が失効しました	EDCT を取り消します
start up　(report ready to taxi)	
エンジンを始動して（地上走行準備完了を通報して）下さい	
expect start up at ~	expect departure at ~ or later
エンジン始動予定時刻は～です	出発は～以後の予定です
expect departure ~ minutes behind ~	expect departure after arrival of ~
出発は～の～分後の予定です	出発は～着陸後の予定です

＊（*1）～（*4）EDCT の代わりに Expected Departure Clearance Time でもよい．

✈ POINT:　IFR 出発機の動き－1

　IFR で出発する航空機の，離陸までの流れは以下の通りである．
1．出発準備完了　⇒　飛行計画承認要求
2．飛行計画承認伝達　⇒　飛行計画承認の確認（無線若しくは DCL による）
　　　　　　　　　　　　　　　（＊離陸準備完了までに承認を確認）
3．地上走行要求　⇒　地上走行指示　⇒　地上走行開始
4．離陸準備完了　⇒　離陸許可　⇒　離陸

Introduction

　IFR で出発する航空機は原則として移動開始の約 5 分前に管制承認（ATC Clearance）を要求する．管制承認とは，航空機の運航者が通報した飛行計画に基づいてその飛行を大臣が承認するものである（必要に応じて内容変更される場合もある）．なお，飛行計画に記入された経路のみの承認であり，飛行計画の高度を承認するものではない場合がある．

　管制機関（等）に管制承認を要求する場合は通常，

　　　1．目的地

　　　2．巡航予定高度

　　　3．駐機位置

を適宜通報する．

　管制承認は，以下の事項を含み，必要なものがその順に発出される．

　　　1．航空機のコールサイン

　　　2．管制承認限界点（クリアランスリミット）

　　　3．出発方式（SID, transition）

　　　4．飛行経路

　　　5．高度

　　　6．その他必要な事項（出発制限時刻，待機指示，二次レーダーコード，マック数等）

　クリアランスリミットは，目的地まで IFR で飛行する場合は目的飛行場，IFR から VFR への変更や待機が予想される場合等管制上必要な場合はフィックス（P.87 参照）となる．なお，旅客機の場合，クリアランスリミットは目的飛行場としていることが多い．

　しかし，目的飛行場までの管制承認を受領したとしても，着陸のための進入の許可は含んでいない．

　また，EOBT の 3 時間前以降に飛行計画の経路を訂正・変更した場合であって，「via flight planned route」の用語による承認を受けた場合等管制承認の経路に疑義が生じたときは，パイロットは，経路の一部又は全部について管制機関に確認すべきである．

トラフィックが多い飛行場

管制承認		地上走行等		離着陸		上昇
DEL	⇒	GND	⇒	TWR	⇒	DEP/APP
GND	⇒	GND	⇒	TWR	⇒	DEP/APP
TWR	⇒	TWR	⇒	TWR	⇒	DEP/APP/ACC
AFIS	⇒	AFIS	⇒	AFIS	⇒	DEP/APP/ACC

トラフィックが少ない飛行場　　　　　　　　＊なお，フライトサービスは施設によって異なる

Basic Example

飛行場管制所管制承認伝達席（Tokyo Delivery）・飛行場管制所地上管制席（Tokyo Ground）と交信

> **PIL:** Tokyo Delivery, Nippon Air 710, spot 12.
>
> **DEL:** Nippon Air 710, Tokyo Delivery, cleared to Osaka airport via Laxas Three Departure, flight planned route, maintain FL 200, squawk 2320.
>
> **PIL:** Nippon Air 710 cleared to Osaka airport via Laxas Three Departure, flight planned route, maintain FL 200, squawk 2320.
>
> **DEL:** Nippon Air 710, monitor Tokyo Ground, report when ready.
>
> **PIL:** Monitor Tokyo Ground, report when ready, Nippon Air 710.
>
> **PIL:** Tokyo Ground, Nippon Air 710, spot 12, information E, request push back.
>
> **GND:** Nippon Air 710, Tokyo Ground, push back approved, runway 16R, heading south.
>
> **PIL:** Push back approved, runway 16R, heading south, Nippon Air 710.

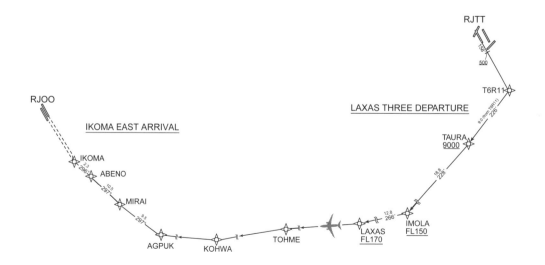

Phraseology Example 1

　管制承認の簡素化（SDC；Simplified Departure Clearance）により，巡航高度の代わりに初期上昇高度によるクリアランスが発出される場合は，以下のようになる.

　なお，管制承認に関わる高度を発出する際，航空交通状況等の理由によりパイロットが要求した高度を管制機関が指定できない場合，原則として指定可能な高度に続き巡航高度として予定する高度が「expect」を前置して通報される.

PIL:　　Kumamoto Ground, Nippon Air 652.

GND:　　Nippon Air 652, Kumamoto Ground, go ahead.

PIL:　　Nippon Air 652, to Tokyo, FL 370, spot 5, we have D.

GND:　　Nippon Air 652 cleared to Tokyo airport via Mifne One Departure, Spide Transition, flight planned route, maintain 6,000, expect FL 370, squawk 3305.

PIL:　　Nippon Air 652 cleared to Tokyo airport via Mifne One Departure, Spide Transition, flight planned route, maintain 6,000, expect FL 370, squawk 3305.

GND:　　Nippon Air 652, read back is correct.

　管制承認は，通常，出発機が「ready」になる間に，他の出発便で同経路・同高度を要求している航空機がない等の交通流を考慮・確認してから発出される.

　管制承認伝達業務の簡素化方式が行われている所においては，あらかじめ空港管制機関と管制部が管制承認の内容について調整し（通常，国際線等を除く），逐一行っている空港管制機関と管制部との間の管制承認に関わるやりとりが省略され，空港の出発間隔のみを考慮して航空機へ管制承認が伝達される.

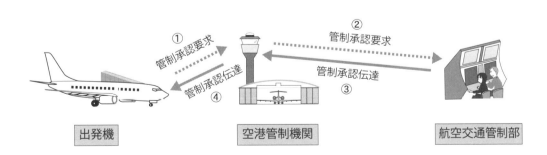

① 管制承認要求　② 管制承認要求　管制承認伝達 ③　管制承認伝達 ④

出発機　　　空港管制機関　　　航空交通管制部

Phraseology Example 2

出発制御時刻（EDCT：Expected Departure Clearance Time）によって，特定の時刻以降に離陸するよういわれる場合もある．多くの場合，EDCT の 1 〜 3 分前から 10 分〜 25 分後の範囲内で離陸する．

PIL:	Fukuoka Delivery, Nippon Air 714.
DEL:	Nippon Air 714, Fukuoka Delivery, go ahead.
PIL:	Nippon Air 714, destination Narita, request FL 390, spot 9.
DEL:	Nippon Air 714 cleared to Narita airport via Yokat Five Departure, Sabar Transition, flight planned route, maintain FL 170, expect FL 390, squawk 2046, EDCT 0635.
PIL:	Nippon Air 714 cleared to Narita airport via Yokat Five Departure, Sabar Transition, flight planned route, maintain FL 170, expect FL 390, squawk 2046, EDCT 0635.
DEL:	Nippon Air 714, read back is correct, Ground frequency 121.7.
PIL:	121.7, Nippon Air 714.
PIL:	Fukuoka Ground, Nippon Air 714, spot 9, request push back with Q.
GND:	Nippon Air 714, Fukuoka Ground, push back approved, runway 34.
PIL:	Push back approved, runway 34, Nippon Air 714.

航空交通管理センター（ATMC）により航空交通流管理のためフローコントロールが実施される場合は，必要に応じて「due to flow control」の用語が付され出発時刻の制限を受けることがあり，ノータム（RJJW / RJJX / RJJY / RJJZ）によってその実施状況が提供される．

Phraseology Example 3

　管制圏のある飛行場では，パイロットに対して管制官から「cleared to」という用語を用いて直接発出されるが，管制官のいない飛行場（飛行場対空援助業務が実施される空港）においては，「ATC clears」が前置され，ACC からの情報が運航情報官経由でパイロットに伝達される．

PIL:　　Fukushima Radio, Nippon Air 82.

AFIS:　Nippon Air 82, Fukushima Radio, go ahead.

PIL:　　Nippon Air 82, to Osaka, proposing FL 340, gate 3.

AFIS:　Nippon Air 82, roger, to Osaka, FL 340, temperature 17, QNH 2988, stand by for clearance.

PIL:　　QNH 2988, stand by for clearance, Nippon Air 82.

AFIS:　Nippon Air 82, clearance.

PIL:　　Nippon Air 82, go ahead.

AFIS:　ATC clears Nippon Air 82 cleared to Osaka airport via Nasno Three Departure, flight planned route, maintain 9,000, expect FL 340, squawk 2461, hold on the ground.

PIL:　　ATC clears Nippon Air 82 cleared to Osaka airport via Nasno Three Departure, flight planned route, maintain 9,000, expect FL 340, squawk 2461, hold on the ground.

AFIS:　Nippon Air 82, read back is correct.

PIL:　　Fukushima Radio, Nippon Air 82, commenced push back.

AFIS:　Nippon Air 82, roger, push back, runway 01.

　管制間隔設定のための出発制限解除時刻の指定がない場合は，地上待機の指示「hold on the ground」の用語が使用され，その指示が取り消される場合は，「released for departure」（出発制限を解除します）の用語が使用される（P.20 参照）．出発制限解除時刻が指定される場合は「released at ~」（出発制限を～に解除します）が使用される．なお，「~ expect FL 340, squawk 2461, clearance void if not off the ground by 0350 (vifno 0350)」のように管制承認失効時刻（clearance void time）が付される場合もある．この場合は，その時刻までに離陸しないと管制承認は無効となる．

Phraseology Example 4

　フライトプランと異なる高度を要求する場合は，以下のように，「altitude change」の用語等をつければよい．

PIL:	Tottori Radio, Nippon Air 290.
AFIS:	Nippon Air 290, Tottori Radio, go ahead.
PIL:	Nippon Air 290, Tokyo airport, spot 2, and request altitude change, propose FL 290.
AFIS:	Nippon Air 290, to Tokyo, FL 290, roger, stand by for clearance.
PIL:	Stand by clearance, Nippon Air 290.
AFIS:	Nippon Air 290, Tottori Radio, clearance.
PIL:	Nippon Air 290, go ahead.
AFIS:	ATC clears Nippon Air 290 cleared to Tokyo airport via Miyazu One Departure, flight planned route, maintain 7,000, expect FL 290, squawk 1775.
PIL:	ATC clears Nippon Air 290 cleared to Tokyo airport via Miyazu One Departure, flight planned route, maintain 7,000, expect FL 290, squawk 1775.
AFIS:	Nippon Air 290, read back is correct, report commence push back.
PIL:	Report commence push back, Nippon Air 290.
PIL:	Tottori Radio, Nippon Air 290, commence push back, runway 28.
AFIS:	Nippon Air 290, push back runway 28, wind 160 degrees at 5 knots, temperature 24, QNH 2959 inches, report commence taxi.
PIL:	2959, report commence taxi, Nippon Air 290.

Phraseology Example 5

レーダー管制が行われていない飛行場対空援助業務が実施されている空港等では，SID 又は Transition の経路を省略して航路上のフィックスへ直行する場合もある．この場合，VMC を維持して上昇することを条件に承認される場合がある．これは「climb in VMC」と呼ばれる．

AFIS: Nippon Air 2232, clearance.

PIL: Nippon Air 2232, go ahead.

AFIS: ATC clears Nippon Air 2232 cleared to Osaka airport via direct Niigata, flight planned route, maintain 10,000, expect FL 280, climb in VMC until 10,000, squawk 3451.

PIL: ATC clears Nippon Air 2232 cleared to Osaka airport, after airborne direct Niigata, flight planned route, maintain 10,000, expect FL 280, climb in VMC until 10,000, squawk 3451.

AFIS: Nippon Air 2232, read back is correct.

PIL: Yamagata Radio, Nippon Air 2232, request taxi.

AFIS: Nippon Air 2232, radio advises hold short of runway, traffic short on final, touch and go.

「climb in VMC」の要求によって，以下のようなことが期待できる．

1．経路上のフィックスへの直行

SID 等の省略

2．飛行経路（SID 等）の高度制限の解除

出発経路上の CB 等を避けるために，VMC を維持して航空路に合流できるような場合

3．到着機による出発制限の回避

他の IFR 機のために待機を指示されたような場合，「climb in VMC」の要求によって，「hold on the ground」が解除されることがある

他機との衝突回避・障害物衝突回避・航法等はすべてパイロットの責任において行わなければならない．また，承認された高度以外の飛行は禁止である．

上記のように，承認経路又は通常承認される飛行経路と異なる経路を VMC を維持して上昇したいパイロットは，管制機関へその旨要求し（request direct ***），承認を得なければならない．

なお，飛行経路上の高度制限の解除を目的とした「climb in VMC」の承認を受けたパイロットは，当該飛行経路に付されている高度制限には拘束されないが，承認された SID の経路については拘束される．

　出発の可否を検討する際，出発飛行場が着陸のための最低気象条件を満たしているかの問題がある．これは離陸直後の不具合等により目的地への飛行を断念した場合に引き返して着陸できるかどうかが問題となるからである．

IFR による出発の可否

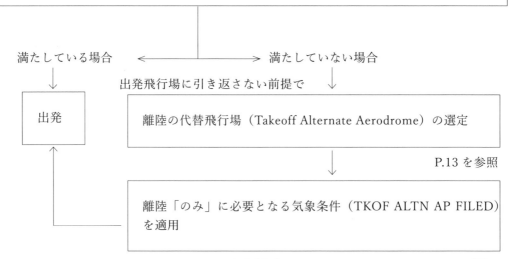

P.13 を参照

P.11 を参照（AIP の TKOF ALTN AP FILED の欄）

（*1）SIDs are designed in accordance with provisional standards for FLIGHT PROCEDURE DESIGN と注記される．注記されていない場合は「飛行方式設定基準」による．なお，飛行方式設定基準における離陸の最低気象条件は，以下の通りである．

CAT II / III	CAT II / III 精密進入の最低気象条件の値に等しい RVR
CAT I	CAT I 精密進入の最低気象条件の値に等しい RVR（使用できない場合は地上視程）
非精密進入	非精密の MDH に等しい雲高（100 ft 単位に切り上げ），及び，最低気象条件の値に等しい RVR（使用できない場合は地上視程）
周回進入	周回進入の MDH に等しい雲高（100 ft 単位に切り上げ），及び，最低気象条件の値に等しい地上視程

　なお，「最低気象条件の値に等しい地上視程」とは，公示された最低気象条件の数値に等しい値である（例：進入方式に対して公示された最低気象条件が CMV 2,000 m であれば，離陸の最低気象条件として地上視程 2,000 m として適用する）．

　通常，「TKOF ALTN AP FILED」の気象条件の方が, 着陸のための最低気象条件 (Weather Minima) より, より厳しい条件での離陸が可能である.

　なお，離陸の最低気象条件は, 離陸滑走中に利用する,

　　　1. 滑走路灯：REDL

　　　2. 滑走路中心線灯：RCLL

　　　3. 滑走路中心線標識：RCL Marking

の設置・運用状況により必要となる地上視程・RVR は異なる.

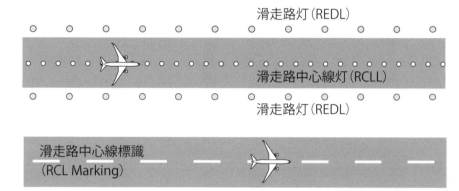

RJEO AD 2.22 FLIGHT PROCEDURES

TAKE OFF MINIMA

	RWY	REDL AVBL	REDL OUT
		CEIL-VIS	CEIL-VIS
TKOF ALTN AP FILED	13	300´-1000m	300´-1200m
	31		
Other	13	AVBL LDG MINIMA	
	31		

NOTE: SIDs are designed in accordance with provisional standards for FLIGHT PROCEDURE DESIGN.

RJFM AD 2.22 FLIGHT PROCEDURES

1. TAKE OFF MINIMA

	RWY	ACFT CAT	REDL & RCLL		REDL or RCLL or RCL Marking		NIL (DAYTIME ONLY)	
			RVR	VIS	RVR	VIS	RVR	VIS
Multi-Engine ACFT with TKOF ALTN AP FILED	09	A,B,C,D	-	400m	-	400m	-	500m
	27	A,B,C,D	400m	400m	400m	400m	-	500m
OTHER	09	A,B,C,D	AVBL LDG MINIMA					
	27	A,B,C,D						

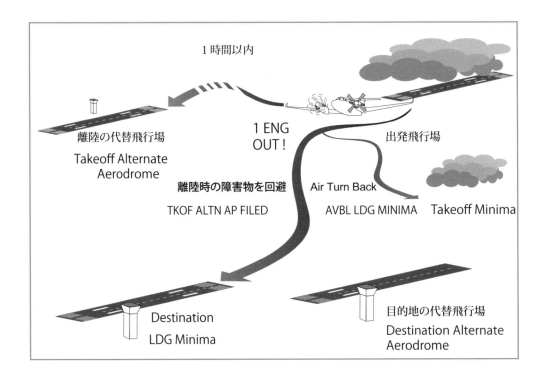

1 時間以内

離陸の代替飛行場
Takeoff Alternate
Aerodrome

1 ENG
OUT !

出発飛行場

離陸時の障害物を回避
TKOF ALTN AP FILED

Air Turn Back
AVBL LDG MINIMA

Takeoff Minima

Destination
LDG Minima

目的地の代替飛行場
Destination Alternate
Aerodrome

✈ POINT:　　目的地に対する代替飛行場 (Destination Alternate Aerodrome)

　　IFR による飛行の場合，飛行を計画する段階で代替飛行場及び搭載燃料量について検討を行う．目的地に対する代替飛行場に関しては以下の通りである．

CAT II / III	CAT I 精密進入の最低気象条件の値に等しい地上視程
CAT I	非精密進入の MDH に等しい雲高（100 ft 単位に切り上げ），及び，最低気象条件の値に等しい地上視程
非精密進入	当該進入方式の MDH に 200 ft を加えた雲高（100 ft 単位に切り上げ），及び，最低気象条件に 1,000 m を加えた地上視程
周回進入	周回進入の MDH に等しい雲高（100 ft 単位に切り上げ），及び，最低気象条件に等しい地上視程

POINT:　離陸の代替飛行場（Takeoff Alternate Aerodrome）の選定

　出発飛行場へ引き返さない場合，緊急の事態に備え近辺に着陸できる飛行場を選定しておく必要がある．この飛行場を「Takeoff Alternate Airport」という．なお，必要に応じてフライトプランに記入する．

１．離陸の代替飛行場：離陸時の気象状況が着陸最低気象条件未満の場合，

・単発機は離陸できない．

・多発機は離陸直後に１個の発動機が停止し，無風状態で

・双発機は１時間以内

・３発以上の航空機は２時間以内

に到達できる範囲内に代替飛行場としての最低気象条件を満足した代替飛行場を選定すればよい．

２．離陸の代替飛行場としての最低気象条件

CAT II / III	CAT II / III 精密進入の最低気象条件の値に等しい地上視程
CAT I	CAT I 精密進入の最低気象条件の値に等しい地上視程
非精密進入	非精密進入の MDH に等しい雲高（100 ft 単位に切り上げ），及び，最低気象条件の値に等しい地上視程
周回進入	周回進入の MDH に等しい雲高（100 ft 単位に切り上げ），及び，最低気象条件の値に等しい地上視程

UNIT.2. Taxiing & Take-off

Words & Phrases

cross runway ~	backtrack runway ~
滑走路〜の横断を許可します	滑走路〜をバックトラックして下さい
taxi via runway ~	continue taxiing
滑走路〜を地上走行して下さい	地上走行を続けて下さい
remain this frequency	change to my frequency ~
この周波数にとどまって下さい	〜に変更して下さい

Introduction

　航空機が地上走行を開始するときは，自機の現在位置を通報してクリアランスを得なければならない．走行経路については，指示が省略された場合，別に取り決めがある場合を除いては任意の経路を走行することができる．

　なお，パイロットは気象状態が離陸のステートミニマ又は自分に適用される離陸の最低気象条件を満たさない場合は，その旨を通報し，離陸してはならない．離陸の最低気象条件を満たさない旨の通報を行わなかった場合は，管制間隔が設定され次第，気象状態に関わらず離陸許可が発出される可能性がある．

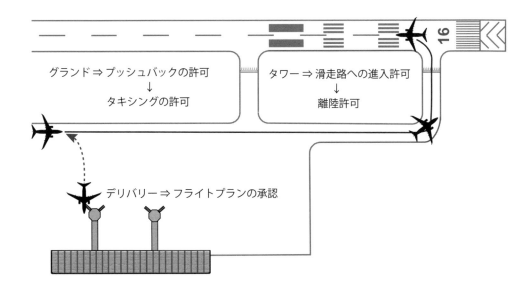

Basic Example

飛行場管制所地上管制席（Tokyo Ground）・飛行場管制所（Tokyo Tower）と交信

PIL:　Tokyo Ground, Nippon Air 710, spot 12, request taxi.

GND:　Nippon Air 710, runway 16R, taxi via W-6, A, to holding point A-14.

PIL:　Runway 16R, taxi via W-6, A, to holding point A-14, Nippon Air 710.

GND:　Nippon Air 710, contact Tower 118.1.

PIL:　Contact Tower 118.1, Nippon Air 710.

PIL:　Tokyo Tower, Nippon Air 710, ready.

TWR:　Nippon Air 710, Tokyo Tower, runway 16R at A-14, line up and wait.

PIL:　Runway 16R at A-14, line up and wait, Nippon Air 710.

TWR:　Nippon Air 710, wind 150 at 12, runway 16R at A-14, cleared for take-off.

PIL:　Runway 16R at A-14, cleared for take-off, Nippon Air 710.

TWR:　Nippon Air 710, contact Departure.

PIL:　Contact Departure, Nippon Air 710.

Phraseology Example 1

特定の空港では，走行経路が AIP に公示されている標準走行経路の名称によって指示される場合がある．

PIL:　　Tokyo Ground, Nippon Air 710, spot 7, request push back, information X.

GND:　　Nippon Air 710, Tokyo Ground, push back approved, runway 05.

PIL:　　Nippon Air 710, push back, runway 05.

PIL:　　Tokyo Ground, Nippon Air 710, request taxi.

GND:　　Nippon Air 710, runway 05, taxi via W, route 5.

PIL:　　Runway 05, taxi via W, route 5, Nippon Air 710.

Route ID	Direction	Routing Via
Route 5	To RWY05	TWY A-R-S
	From RWY23	TWY S-R-A

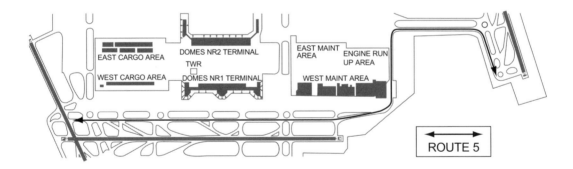

ROUTE 5

✈ POINT:　Runway Safety

　滑走路の安全（runway safety）には，以下の３つのキーワードがある．

1．滑走路誤進入（runway incursion）

2．滑走路逸脱（runway excursion）

3．滑走路誤認識（runway confusion）

Phraseology Example 2

使用滑走路へ向かう途中又はスポットへ向かう途中に，他の滑走路の横断を必要とする場合は，航空機が当該滑走路に近づいたときに，「cross runway」か「hold short of runway」が管制官から指示される．勝手に滑走路を横断してはならない．滑走路誤進入を防止するためにも，この指示に対してパイロットは必ずリードバックしなければならない．

PIL:	Tokyo Tower, Nippon Air 163, spot 56, approaching L-3.
TWR:	Nippon Air 163, Tokyo Tower, hold short of runway 16R due to departure traffic.
PIL:	Hold short of runway 16R, Nippon Air 163.
TWR:	Nippon Air 163, cross runway 16R at L-3, contact Ground 121.7.

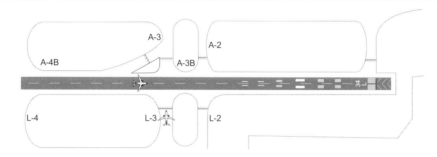

Phraseology Example 3

インターセクション・デパーチャーを行う場合は，パイロットはインターセクション名をつけて要求する．

PIL:	Kumamoto Tower, Nippon Air 652, ready, request T-2 intersection departure.
TWR:	Nippon Air 652, Kumamoto Tower, T-2 intersection approved, runway 07 at T-2, line up and wait.
PIL:	T-2 intersection approved, runway 07 at T-2, line up and wait, Nippon Air 652.
TWR:	Nippon Air 652, wind 310 at 3, runway 07 cleared for take-off.

Phraseology Example 4

離陸滑走前に離陸後のヘディングが指示された場合は，離陸後速やかに指示されたヘディングに旋回する．

PIL:	Tokyo Tower, Nippon Air 710, ready.
TWR:	Nippon Air 710, Tokyo Tower, after airborne, turn right heading 120, wind 080 at 12, runway 05 cleared for take-off.
PIL:	After airborne, heading 120, runway 05 cleared for take-off, Nippon Air 710.

なお，「continue runway heading」の用語が用いられる場合があるが，「runway heading」とは，「滑走路中心線の磁方位のヘディング」を意味している（磁方位が公示されていない滑走路では，滑走路上にラインアップしたときのヘディングを維持して飛行）．

一方，SID における「climb runway heading」(Climb RWY HDG / Climb via RWY HDG 等) は，滑走路中心線の延長線を飛行する（風の影響がある場合等は，WCA をとって上昇する）必要がある．

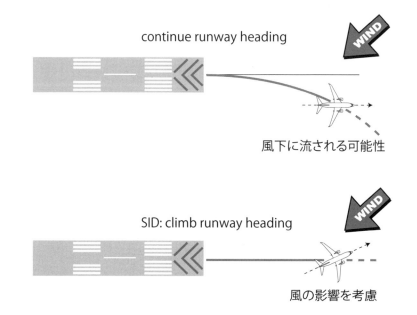

continue runway heading

風下に流される可能性

SID: climb runway heading

風の影響を考慮

Phraseology Example 5

　グライドパス停止線（グライドスロープの電波障害を防止するために航空機を待機させる目的で設置された停止線）が設けられている空港では，タクシー中の航空機は管制官から別途指示を得なければそれを通過してはならない．

PIL:　　Fukuoka Ground, Nippon Air 714, request taxi.

GND:　　Nippon Air 714, runway 34, taxi via Y, A, to holding point E-13, cross GP hold line.

PIL:　　Runway 34, taxi via Y, A, to holding point E-13, cross GP hold line, Nippon Air 714.

GND:　　Nippon Air 714, contact Tower 118.4.

PIL:　　Contact Tower 118.4, Nippon Air 714.

PIL:　　Fukuoka Tower, Nippon Air 714, ready.

TWR:　　Nippon Air 714, Fukuoka Tower, hold short of runway 34.

PIL:　　Hold short of runway 34, Nippon Air 714.

TWR:　　Nippon Air 714, runway 34 line up and wait, arrival approaching base.

PIL:　　Runway 34 line up and wait, Nippon Air 714.

TWR:　　Nippon Air 714, arrival turning base, wind 310 at 8, runway 34 cleared for take-off.

　気象状態が雲高 800 ft 以上，かつ地上視程 3,200 m 以上であって，ILS 進入方式により進入する到着機がある場合は，航空機にグライドパス停止線の通過を指示した後，ILS 進入方式により進入を開始した到着機に対して，「glide slope signal not protected」の用語により，グライドスロープの電波精度が確保されていない旨が通報される．

　気象状態が，雲高 800 ft 未満又は地上視程 3,200 m 未満であって，ILS 進入方式により進入する到着機がアプローチゲート（P. 63 参照）を通過した場合は，管制機関は停止線の通過を指示しない．ただし，到着機が滑走路の視認を通報した場合は，当該機に対してグライドスロープの電波精度が確保されていない旨が通報される．

　なお，電波に障害を与えないことが検証された型式の航空機については，停止の必要はない．

Phraseology Example 6

　飛行場対空援助業務が実施されている空港では,「cleared for take-off」の用語は使用されず,「runway is clear」の用語が使用される. なお,「traffic not reported (in the vicinity of this ~ airport)」とは, 空港周辺に報告されたトラフィックがないことを示す用語である.

PIL: 　　Fukushima Radio, Nippon Air 82, request taxi information.

AFIS:　 Nippon Air 82, taxi to holding point runway 01, radio advises hold short of runway 01, inbound Boeing 737 7 miles on final.

PIL: 　　Taxi to holding point runway 01, hold short of runway 01, Nippon Air 82.

PIL: 　　Nippon Air 82, ready.

AFIS:　 Nippon Air 82, radio advises hold short of runway 01.

PIL: 　　Hold short of runway 01, Nippon Air 82.

AFIS:　 Nippon Air 82, clearance.

PIL: 　　Go ahead.

AFIS:　 ATC clears Nippon Air 82 released for departure.

PIL: 　　Nippon Air 82, released for departure.

AFIS:　 Nippon Air 82, read back is correct, wind 230 at 3 knots, runway 01 runway is clear.

PIL: 　　Runway 01 runway is clear, Nippon Air 82.

AFIS:　 Nippon Air 82, contact Tokyo Control 128.2.

　上記会話のように, 周囲の状況から, パイロットに対して必要な措置が助言される場合,「radio advises」の用語が使用されることがある.

　なお, EDCT と管制機関による管制間隔設定のため出発制限（解除時刻）が同一の航空機に重複して指定される場合がある. この場合, 当該機には各々の制限による出発可能な時刻のいずれか遅い時刻が適用される.

　仮に, EDCT の時刻よりも前に「released for departure」が伝達された場合でも,（「hold on the ground」の解除のみを意味するので）EDCT は有効である.

Phraseology Example 7

　飛行場対空援助業務が実施されている空港では，一般的には，プッシュバック及びタクシーの開始は，周囲の状況を勘案したパイロットの判断に任されており，その行動の開始を通報すればよい．なお，滑走路使用方向と逆方向への地上走行を行う場合には，「backtrack」の用語が使用される．

PIL:　　 Tottori Radio, Nippon Air 290, commence taxi, runway 28.

AFIS:　 Nippon Air 290, backtrack runway 28, surface wind 150 at 6 knots.

PIL:　　 Backtrack runway 28, Nippon Air 290.

AFIS:　 Nippon Air 290, Tottori Radio.

PIL:　　 Nippon Air 290, ready, go ahead.

AFIS:　 Nippon Air 290, new QNH 2957, and cumulonimbus 15 kilometers south, moving to northeast, wind 150 at 6 knots, runway 28 runway is clear.

PIL:　　 Runway 28 runway is clear, 2957, Nippon Air 290.

PIL:　　 Tottori Radio, Nippon Air 290, airborne 03.

AFIS:　 Nippon Air 290, roger, contact Tokyo Control 133.8.

PIL:　　 Tokyo Control 133.8, Nippon Air 290.

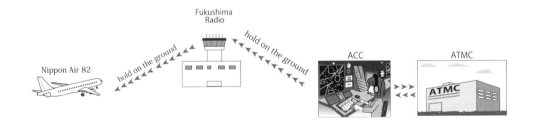

EDCT と管制間隔設定のため出発制限解除時刻

□ 「ATC clears *** cleared to ~ airport via South One Departure, flight planned route, maintain 9,000, expect FL 300, squawk 3653, EDCT 0840」

□ 「ATC clears *** hold on the ground」

□ 「request taxi → taxi to holding point runway 01, and ATC clears released for departure」

　　　→ EDCT は依然として有効である→航空機は EDCT まで離陸を待つ

UNIT.3. Climb

Words & Phrases

revised clearance	recleared ~
管制承認を変更します	~を承認します
after passing ~ , climb (descend) and maintain ~	
~通過後に上昇（降下）して~を維持して下さい	
turn right (left) heading ~ (*1)	fly heading ~
右（左）旋回針路~	針路~を飛行して下さい
resume own navigation	continue runway heading (*2)
通常航法に戻って下さい	滑走路の方位で飛行して下さい
do not exceed ~ knots	
~ノットを超えて加速しないよう飛行して下さい	
ident for position confirmation	
位置を確認するためにアイデントを送って下さい	
ident observed	
アイデント観察しました	

＊（*1）レーダー誘導は離陸直後の誘導を除き原則として MVA 以上の高度で，ヘディングの指示によって行われ，指示されるヘディングは常に磁方位である（現針路が不明で，かつ，それを確認する余裕がない場合，旋回の度数及び旋回方向が指定される）．また，無線施設が停波した場合（4 時間程度以内），レーダー管制が実施されていれば，レーダー誘導によるクリアランスが発出される．

＊（*2）航空機は使用する滑走路の磁方位の磁針路で飛行する．この場合，偏流の修正を行わない磁針路であることに注意する．（P.18 参照）

Introduction

　レーダー業務が行われている飛行場では，通常，滑走路終端から１マイル以内でレーダーターゲットが捕捉され「radar contact」の用語とともに直ちにレーダーサービスが開始される．これ以外の場合は，トランスポンダーの操作等を求められる場合もある．

　ターミナルレーダー管制業務が行われていない飛行場では，ACC と通信設定を行った後にレーダー覆域内でレーダー識別され，「radar contact」の用語とともに現在位置が確認される場合もある．

　パイロットはレーダー誘導が開始されない限りは SID を飛行することとなる．離陸後，レーダー誘導が開始された場合は，承認された SID に関わらず誘導の指示が優先する．

　レーダー誘導は，ヘディングを指定する用語によって開始され，それに引き続き誘導目的・目標が通報される（ただし，目的又は目標の一方の通報により他方が明白な場合は，いずれかが通報される）．SID，トランジション（又は STAR）による飛行の場合，レーダー誘導，フィックスへの直行等，飛行中に経路の変更が指示された場合（その他，指定高度の変更・レーダー誘導の終了を含む）は，必要となる高度制限についてあらためて指示される（言及されなければ，変更後の経路に付加されている高度制限は，SID（STAR 等）に公示されたものを含めて，すべてキャンセルされる）．

　新たな管制機関への通信移管が行われた場合，パイロットは上昇中であれば通過高度（100 ft 単位）と指定高度を通報し，巡航中であれば指定高度を通報する．

✈ POINT:　IFR 出発機の動き−２

　IFR で出発する航空機は，通常，以下のようなルートをたどる．
1．滑走路から離陸　⇒　SID　⇒　航空路等
2．滑走路から離陸　⇒　SID　⇒　Transition ⇒　航空路等
3．滑走路から離陸　⇒　レーダー誘導　⇒　航空路等
4．滑走路から離陸　⇒　SID　⇒　レーダー誘導　⇒　航空路等
5．滑走路から離陸　⇒　レーダー誘導　⇒　SID　⇒　航空路等

Basic Example

ターミナル管制所出域管制席（Tokyo Departure）と交信

> **PIL:** Tokyo Departure, Nippon Air 710, leaving 1,800 climbing FL 200.
> **DEP:** Nippon Air 710, Tokyo Departure, radar contact.
>
> **DEP:** Nippon Air 710, contact Tokyo Control 120.5.
> **PIL:** Contact Tokyo Control 120.5, Nippon Air 710.

Phraseology Example 1

SID に設定された高度制限がすべて有効な場合は，以下のようになる．

> PIL: Tokyo Departure, Nippon Air 710, leaving 1,200, FL 200.
>
> DEP: Nippon Air 710, Tokyo Departure, radar contact.

　上記の場合，SID 等に設定された高度制限を保った後，承認されている高度（初期上昇高度）である FL 200 まで上昇することができるが，以下の場合においては，承認されている高度である FL 200 への上昇には新たな指示が必要である．

> PIL: Tokyo Departure, Nippon Air 710, leaving 1,200, FL 200.
>
> DEP: Nippon Air 710, Tokyo Departure, radar contact, climb and maintain 9,000.

　下記の場合は，承認されている高度である FL 200 を指定されているため，途中の高度制限に関係なく，高度 FL 200 まで制限なく上昇する．

> PIL: Tokyo Departure, Nippon Air 710, leaving 1,200, FL 200.
>
> DEP: Nippon Air 710, Tokyo Departure, radar contact, climb and maintain FL 200.

STANDARD DEPARTURE CHART-INSTRUMENT

RJTT/TOKYO INTL   RNAV SID

LAXAS THREE DEPARTURE	RNAV1

Note 1) DME/DME/IRU or GNSS required. ※The aircraft equipped with only DME/DME/IRU must be able to update its position without delay at the starting point of take-off rolling. 2) RADAR service required.		Critical DME	RWY16R: HME 1.2NM FM DER - 1.9NM to T6R11 HYD T6R11 - TAURA RWY16L: HME 1.0NM FM DER - 2.4NM to T6L21 HYD 9.0NM to TAURA - TAURA RWY34R: HME 1.0NM FM DER - 2.5NM to TT502 HYD 8.6NM to TAURA - TAURA RWY34L: HME 0.5NM FM DER - 2.5NM to TT502 HYD 8.6NM to TAURA - TAURA RWY04: HME 1.7NM FM DER - 2.5NM to TT502 HYD 8.6NM to TAURA - TAURA RWY05: HME DER - 2.7NM to TT502 HYD 8.6NM to TAURA - TAURA
DME GAP	RWY16R:DER - 1.2NM FM DER RWY16L:DER - 1.0NM FM DER RWY34R:DER - 1.0NM FM DER RWY34L:DER - 0.5NM FM DER RWY04:DER - 1.7NM FM DER RWY22:DER - 1.4NM FM DER		
Inappropriate Navaids	See AD1.1.6.10.3.Inappropriate NAVAIDs for RNAV1		

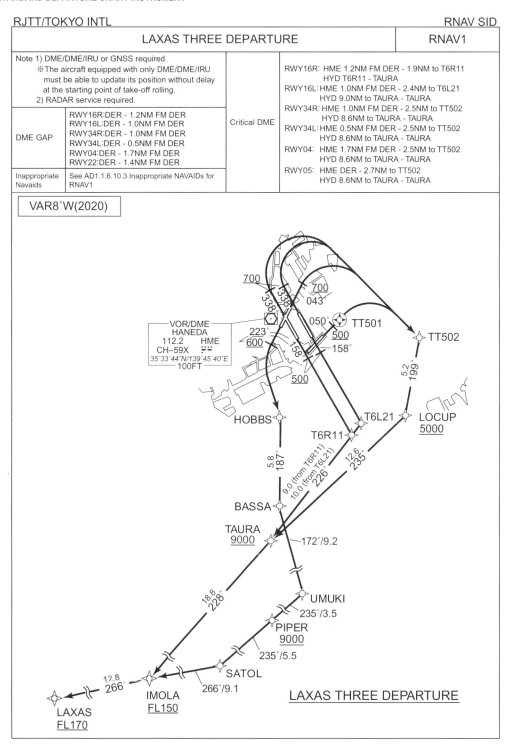

VAR8°W(2020)

LAXAS THREE DEPARTURE

Phraseology Example 2

　SID，トランジション（又はSTAR）による飛行の場合，飛行中においてあらためて高度（現在指定されている高度を含む）が指定される場合，又は，フィックスへの直行を含め飛行経路が変更された場合であって，誘導が終了したのち SID，トランジション（又はSTAR）に公示された高度制限又は速度に従って飛行するよう指示されるときは，「climb via SID to ~」（SIDの制限に従い~まで上昇して下さい）の用語が使用される（ない場合は，すべて無効である）．なお，「climb via SID to ~」の指示後，速やかに上昇を開始する必要がある．

　離陸後，レーダー誘導が開始される場合は，以下のようになる．なお，レーダー誘導終了時の「resume own navigation」は，単に平面上の経路承認を意味し，高度の変更を含むものではないので，新規の高度承認，又は高度変更の要求を行わない限りは最後に指定された高度を維持しなければならない．

PIL:　　Tokyo Departure, Nippon Air 710, leaving 1,700, FL 200.

DEP:　　Nippon Air 710, Tokyo Departure, radar contact, fly heading 150 vector to LAXAS.

PIL:　　Heading 150, Nippon Air 710.

DEP:　　Nippon Air 710, turn right heading 240.

PIL:　　Right heading 240, Nippon Air 710.

DEP:　　Nippon Air 710, resume own navigation direct LAXAS, climb via SID to FL 200.

PIL:　　Direct LAXAS, climb via SID to FL 200, Nippon Air 710.

DEP:　　Nippon Air 710, contact Tokyo Control 120.5.

✈ POINT:　レーダーサービス

　レーダーサービスの１つにレーダー誘導がある．（P.124 参照）

1．レーダーサービスの開始　⇒　radar contact / radar contact + position

　　レーダー誘導の開始　⇒　turn left heading ~ vector to ~ / etc.

　　レーダー誘導の終了　⇒　resume own navigation / etc.

2．レーダーサービスの終了　⇒　radar contact lost / radar service terminated / etc.

レーダー誘導により通過しなかったフィックスに関わる高度制限・速度は適用されないが、SID に合流する指示が出された後に「climb via SID to ~」と指示されれば，SID の合流点以降の高度制限又は速度は有効になる．

なお，下記のような経路の変更の場合も同様である．

PIL:	Tokyo Departure, Nippon Air 710, leaving 1,200 for FL 200.
DEP:	Nippon Air 710, Tokyo Departure, radar contact.
PIL:	Nippon Air 710.
DEP:	Nippon Air 710, recleared direct LAXAS, climb via SID to FL 200.
PIL:	Recleared direct LAXAS, climb via SID to FL 200, Nippon Air 710.
DEP:	Nippon Air 710, contact Tokyo Control 120.5.

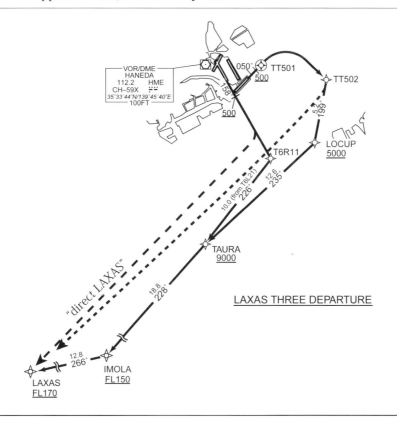

LAXAS THREE DEPARTURE

✈ POINT: 速度制限

　航空法により，通常は以下の制限速度が適用される．

管制圏及び進入管制区のうち，高度 10,000 ft 以下の空域　⇒　250 ノット以下

Phraseology Example 3

　Phraseology Example 2 以外の場合であって，飛行中において，あらためて高度（現在指定されている高度を含む）が指定される場合であって，必要な高度制限についてあらためて指示されるときは，下記のように「climb / descend and maintain ~, comply with restrictions」の用語等により指示される（なお，高度制限について指示がない場合はすべて無効となる）．

PIL:	Tokyo Control, Nippon Air 692, passing 12,300 for FL 150.
ACC:	Nippon Air 692, Tokyo Control, climb and maintain FL 230.
PIL:	Climb and maintain FL 230, Nippon Air 692.
ACC:	Nippon Air 692, cross MOMOT at or above FL 210 due to traffic.
PIL:	Cross MOMOT at or above FL 210, Nippon Air 692.
ACC:	Nippon Air 692, climb and maintain FL 350, comply with restrictions.
PIL:	Climb and maintain FL 350, comply with restrictions, Nippon Air 692.
ACC:	Nippon Air 692, contact Tokyo Control 133.55.

　上記の場合，変更指示（climb and maintain FL 350）前に高度制限（cross MOMOT at or above FL 210）があるため，その高度制限（管制官による指示である cross MOMOT at or above FL 210；その高度制限が SID，トランジション又は STAR に公示されたものではない場合）を守りながら FL 350 に上昇することとなる．

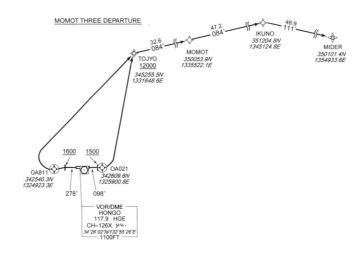

また，レーダー誘導・フィックスへの直行等，飛行経路が変更される場合も同様である．

PIL: Tokyo Departure, Nippon Air 713, leaving 2,000 climbing FL 240.

DEP: Nippon Air 713, Tokyo Departure, radar contact.

PIL: Nippon Air 713.

DEP: Nippon Air 713, turn right heading 190 vector to PIGOK, maintain FL 240.

PIL: Turn right heading 190, maintain FL 240, Nippon Air 713.

DEP: Nippon Air 713, turn right heading 210.

PIL: Turn right heading 210, Nippon Air 713.

DEP: Nippon Air 713, turn right heading 240.

PIL: Turn right heading 240, Nippon Air 713.

DEP: Nippon Air 713, resume own navigation direct PIGOK, comply with restrictions.

PIL: Resume own navigation direct PIGOK, comply with restrictions, Nippon Air 713.

DEP: Nippon Air 713, contact Tokyo Control 120.5.

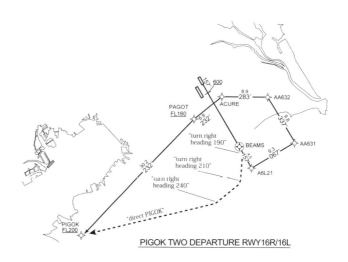

PIGOK TWO DEPARTURE RWY16R/16L

Phraseology Example 4

　レーダー識別された航空機は，その終了までレーダーサービスが継続される．つまり，DEP ⇒ ACC, ACC ⇒ ACC へとレーダーハンドオフが行われても，特に必要がない限りは，「トランスポンダーの操作指示」や「radar contact」の通報等は行われない．

PIL:　　Kumamoto Departure, Nippon Air 652, leaving 3,200 climbing 6,000.

DEP:　　Nippon Air 652, Kumamoto Departure, radar contact, fly heading 230 vector to IWATO, climb and maintain FL 150.

PIL:　　Fly heading 230, climb and maintain FL 150, Nippon Air 652.

DEP:　　Nippon Air 652, turn left heading 180.

PIL:　　Turn left heading 180, Nippon Air 652.

DEP:　　Nippon Air 652, resume own navigation direct IWATO.

PIL:　　Direct IWATO, Nippon Air 652.

DEP:　　Nippon Air 652, contact Kobe Control 118.9.

PIL:　　Contact Kobe Control 118.9, Nippon Air 652.

PIL:　　Kobe Control, Nippon Air 652, leaving 10,600 climbing FL 150.

ACC:　　Nippon Air 652, Kobe Control, climb and maintain FL 230.

PIL:　　Climb and maintain FL 230, Nippon Air 652.

ACC:　　Nippon Air 652, contact Kobe Control 135.65.

MIFNE ONE DEPARTURE / SPIDE TRANSITION

...MIFNE - GOKAH - IWATO - DONAR - SPIDE - Y-23...

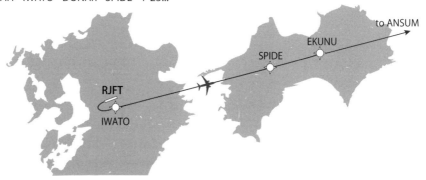

Phraseology Example 5

パイロットの側から VOR やフィックス等への直行を要求することもできる.

> PIL: Fukuoka Departure, Nippon Air 3651, leaving 1,400 for 13,000.
>
> DEP: Nippon Air 3651, Fukuoka Departure, radar contact.
>
> PIL: Nippon Air 3651.
>
> DEP: Nippon Air 3651, fly heading 100 vector to SUOH, report receiving SUOH.
>
> PIL: Heading 100, report receiving SUOH, Nippon Air 3651.
>
> PIL: Fukuoka Departure, Nippon Air 3651, now receiving SUOH, request direct SUOH.
>
> DEP: Nippon Air 3651, resume own navigation direct SUOH.
>
> PIL: Resume own navigation direct SUOH, Nippon Air 3651.
>
> DEP: Nippon Air 3651, contact Tsuiki Approach 119.22.

Phraseology Example 6

クリアランスに「climb in VMC」がある場合, VMC を維持する区間においては, 高度制限に関係なく SID の終点まで飛行すればよい. VMC の維持が困難となった場合は, 直ちにその旨を通報し, 代替のクリアランスを要求又は得る必要がある. その際には, 「unable to maintain VMC (due to ~)」等と通報すればよい.

> PIL: Nippon Air 163, unable to maintain VMC on course, request deviation 10 miles to the east, then proceed direct Kushimoto maintaining VMC.

UNIT.4. En-Route

Words & Phrases

climb (descend) at pilot's discretion maintain ~ (*1)

  パイロットの判断で上昇（降下）して～を維持して下さい

verify at ~	verify assigned altitude ~
～を確認して下さい	指定された高度～を確認して下さい

report leaving / reaching ~

  ～を離脱したら（に到達したら）報告して下さい

report altitude	cruise
高度を知らせて下さい	～でのクルーズを許可します
expect approach at ~ (*2)	expect further clearance at ~ (*3)
進入予定～です	追加管制承認予定～です
delay not determined	no delay expected
遅延時間未定です	遅延の予定ありません
check altimeter setting and confirm (*4)	squawk altitude (*4)
高度計規制値及び *** を確認して下さい	自動高度応答装置を作動させて下さい

stop altitude squawk (*5)

  自動高度応答装置の作動を停止して下さい

verify present altitude	report heading
現在の高度を確認して下さい	針路を知らせて下さい
proceed via last routing cleared	maintain present speed (Mach number)
最後に承認された経路経由で飛行して下さい	現在の速度（マック数）を維持して下さい

＊（*1）この場合，上昇・降下を開始する時機はパイロットの判断に任される．上昇・降下開始後に一時的な水平飛行を行うことはできるが，一度通過した高度に再び降下・上昇してはならない．また，上昇・降下開始後に上昇・降下率の調整を行う場合でも通報しなくてもよい．
＊（*2）進入予定時刻（EAT: Expected Approach Time）とは，到着機が計器進入の許可を得て，進入フィックスを離脱する時刻であって管制官が予想するものをいう．

＊（*3）追加承認予定時刻（EFC: Expect(ed) Further Clearance Time）とは，進入フィックス以外の待機フィックスで待機する航空機が追加承認を得ることができる時刻であって管制官が予想するものをいう．

＊（*4）レーダー画面上に表示された高度と，パイロットの通報した高度との差が 300 ft 以上の場合は確認が行われることがある（300 ft 未満の場合は，表示画面が垂直間隔を設定するために使用される）．なお，自動応答装置の高度は QNE による高度情報である（高度計の表示値ではない）．

＊（*5）上記（*4）の確認を行った後においても 300 ft 以上の高度の差があるときは，自動高度応答装置の作動の中止を指示される場合がある．

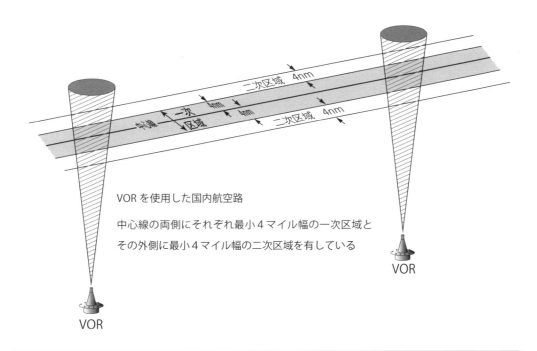

VOR を使用した国内航空路

中心線の両側にそれぞれ最小４マイル幅の一次区域と

その外側に最小４マイル幅の二次区域を有している

✈ POINT:　ATS ルート

　　航空交通業務が実施される飛行経路を ATS ルートと呼ぶ．具体的に以下のものがある．

1．航空路 --- NDB / VOR / TACAN 等，航法無線施設を結んだ経路

2．RNP2 経路 --- RNP2 仕様に従い航行する航空機の用に供するために設定された飛行経路

3．RNAV5 経路 --- RNAV5 仕様に従い航行する航空機の用に供するために設定された経路

4．洋上転移経路 --- 陸上の無線施設と洋上管制区内のフィックスとの間に設定された飛行経路であって，洋上転移経路として公示されたもの

5．直行経路 --- 航空機が無線施設を利用して直行飛行を行うときの飛行経路であって，上記1．〜4．以外のもの

6．SID / Transition / STAR / 等

Introduction

　SID 等により上昇し ATS ルート上（航空路等）のフィックス到達後は，航空機は ATS 経路等を飛行する.

Basic Example

管制区管制所（Tokyo Control）と交信

> PIL:　Tokyo Control, Nippon Air 710, leaving FL 142 for FL 200.
>
> ACC:　Nippon Air 710, Tokyo Control, climb and maintain FL 220.
>
> PIL:　Climb and maintain FL 220, Nippon Air 710.
>
> ACC:　Nippon Air 710, contact Tokyo Control 125.7.
>
> PIL:　Contact Tokyo Control 125.7, Nippon Air 710.
>
> PIL:　Tokyo Control, Nippon Air 710, FL 220.
>
> ACC:　Nippon Air 710, Tokyo Control, roger.
>
> PIL:　Tokyo Control, Nippon Air 710, request descent.
>
> ACC:　Nippon Air 710, descend and maintain FL 200.
>
> PIL:　Descend FL 200, Nippon Air 710.
>
> ACC:　Nippon Air 710, descend and maintain FL 150, expect OHDAI 13,000.
>
> PIL:　Descend FL 150, Nippon Air 710.
>
> ACC:　Nippon Air 710, descend to reach 13,000 by OHDAI, cross OHDAI at 13,000, area QNH 2996.
>
> PIL:　Descend to reach 13,000 by OHDAI, cross OHDAI at 13,000, area QNH 2996, Nippon Air 710.
>
> ACC:　Nippon Air 710, contact Kansai Approach 120.45.
>
> PIL:　Kansai Approach 120.45, Nippon Air 710.

Phraseology Example 1

パイロットは経路の変更を要求したり，又は，管制機関から経路の変更を指示されることがある．

PIL:	Tokyo Control, Nippon Air 710, leaving FL 153 for FL 200.
ACC:	Nippon Air 710, Tokyo Control, climb and maintain FL 220, recleared direct IBENO.
PIL:	Recleared direct IBENO, climb and maintain FL 220, Nippon Air 710.
ACC:	Nippon Air 710, contact Tokyo Control 125.7.

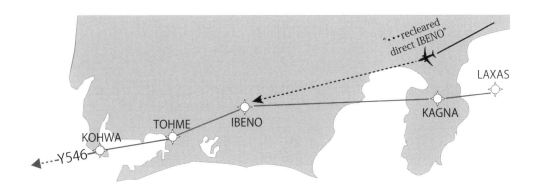

Phraseology Example 2

　ATS 経路等を飛行しているときは，やむを得ない場合を除きその中心線を飛行しなければならず，飛行経路を逸脱する場合は，承認を得なければならない．

　一般的に，悪天（積乱雲等）を避けるためには，左右に避ける，上昇・降下する等が考えられる．ヘディングによる要求，又は，通常航法による回避の場合（deviation による要求の場合）がある．なお，その際には，飛行する方向及び距離も併せて通報することが望ましい．

　deviation（先に承認されていた経路の中心線からの逸脱距離を意味する）による要求の場合は，以下のようになる．なお，パイロットは承認された範囲内で任意の経路を飛行することができる．

PIL:	Tokyo Control, Nippon Air 710, leaving 13,600 for FL 200.
ACC:	Nippon Air 710, Tokyo Control, climb and maintain FL 220.
PIL:	Climb and maintain FL 220, Nippon Air 710.
PIL:	Tokyo Control, Nippon Air 710, request weather deviation, right side, within 10 miles due to weather.
ACC:	Nippon Air 710, roger, deviation 10 miles right approved.
PIL:	Deviation 10 miles right approved, Nippon Air 710.
PIL:	Tokyo Control, Nippon Air 710, now clear of weather.
ACC:	Nippon Air 710, recleared direct IBENO.
PIL:	Recleared direct IBENO, Nippon Air 710.
ACC:	Nippon Air 710, contact Tokyo Control 125.7.

　下記のようにいうこともある．（right / left = romeo / lima）

「request deviation flight 10 miles left of track due to weather」

「request deviation lima side within 10 miles to avoid weather area」

「request deviation right side, south side, within 10 miles due to cloud」

「request weather deviation up to 10 miles left side of route to avoid TCu」

　なお，必要がなくなった場合は，「clear of weather」のような言い方で，その旨を通報することが望ましい．

ヘディングによって要求する場合もある．なお，気象状況等の理由により巡航高度を変更したい場合，下記のように「all the way」や「as final (altitude)」という用語が用いられる場合もある．

PIL:	Tokyo Control, Nippon Air 710, request FL 240 all the way due to light turbulence at FL 220.
ACC:	Nippon Air 710, FL 240 is occupied, request intention.
PIL:	Nippon Air 710, request FL 200.
ACC:	Nippon Air 710, descend and maintain FL 200.
PIL:	Descend and maintain FL 200, Nippon Air 710.
PIL:	Tokyo Control, Nippon Air 710, request heading 250 due to cloud.
ACC:	Nippon Air 710, fly heading 250, report clear of weather.
PIL:	Fly heading 250, report clear, Nippon Air 710.
PIL:	Tokyo Control, Nippon Air 710, request heading 260.
ACC:	Nippon Air 710, fly heading 260.
PIL:	Fly heading 260, Nippon Air 710.
PIL:	Tokyo Control, Nippon Air 710, clear of weather, request direct IBENO.
ACC:	Nippon Air 710, resume own navigation direct IBENO.

交通状況等の理由で，以下のように，ヘディングで飛行する距離を尋ねられる場合もある．

ACC:	Nippon Air 710, how many miles do you request / how many miles would you like to go on this heading.
PIL:	30 miles, Nippon Air 710.

管制機関から示唆された高度が受け入れられない場合は，その旨を通報する．

PIL:	Tokyo Control, Nippon Air 652, request climb FL 390, FL 370 light minus turbulence.
ACC:	Nippon Air 652, how about FL 410.
PIL:	Nippon Air 652, due to performance, unable, request FL 390.
ACC:	Nippon Air 652, roger, stand by.

Phraseology Example 3

エンルート時の降下方法は，大まかにいって，「通常の降下」と「at pilot's discretion の降下」の２つがある．前者の降下指示（「descend and maintain」）に関しては，速やかに通常の降下が行われることが想定されている．他方，後者の「at pilot's discretion の降下」の場合は，降下の開始がパイロットに任せられている．

更に，高度制限を指示する場合であって，特定のフィックスや地点を特定高度で通過する場合等の特別な対応が必要な場合（「descend to reach *** by ***」）の用語もある．

PIL:　　Kobe Control, Nippon Air 619, request descent.

ACC:　　Nippon Air 619, descend and maintain FL 200.

PIL:　　Descend FL 200, Nippon Air 619.

PIL:　　Kobe Control, Nippon Air 619, request direct SIROK.

ACC:　　Nippon Air 619, recleared direct SIROK.

PIL:　　Direct SIROK, Nippon Air 619.

ACC:　　Nippon Air 619, descend to reach 12,000 by SIROK, area QNH 3023.

PIL:　　Descend to reach 12,000 by SIROK, 3023, Nippon Air 619.

ACC:　　Nippon Air 619, contact Kagoshima Radar 121.4.

その他，降下（上昇）の指示に使用されるものとして，以下の用語等がある．

「cross *** at or above (below) ***」

「after passing ***, climb (descend) and maintain ***」

「maintain *** until ***, then climb (descend) and maintain ***」

descend at pilot's
discretion maintain ～

水平飛行 OK
降下の時機は
パイロットの判断

descend to reach ～ by ～

水平飛行 NG
降下の時機は
パイロットの判断

descend and maintain ～

水平飛行 NG
すみやかに降下

Phraseology Example 4

速度調整が行われる場合，指示された IAS ± 10 ノット若しくはマック数の ± 0.02 の範囲内で飛行しなければならない．通常の速度に戻す指示がある場合，以下のようになる．

PIL: Tokyo Control, Nippon Air 652, FL 370.

ACC: Nippon Air 652, Tokyo Control, report Mach number.

PIL: Nippon Air 652, maintaining Mach point 80.

ACC: Nippon Air 652, roger, maintain Mach point 80 or greater.

PIL: Maintain Mach point 80 or greater, Nippon Air 652.

ACC: Nippon Air 652, descend at pilot's discretion maintain FL 290, transit to 300 knots or greater.

PIL: Descend at pilot's discretion maintain FL 290, transit to 300 knots or greater, Nippon Air 652.

ACC: Nippon Air 652, descend to reach FL 220 by SPENS.

PIL: Descend to reach FL 220 by SPENS, Nippon Air 652.

ACC: Nippon Air 652, resume normal speed, contact Tokyo Approach 119.1.

PIL: Resume normal speed, 119.1, Nippon Air 652.

PIL: Tokyo Approach, Nippon Air 652, leaving FL 257 for FL 220, information H.

APP: Nippon Air 652, Tokyo Approach, runway 34L, cleared via Oshima 1K Arrival, descend via STAR to 13,000.

特定速度又はマック数に増速又は減速が指示される場合は，「increase speed to ~」「reduce speed to ~」が使用される．

IAS について特定量増速又は減速することが指示される場合は，「increase speed by ~」「reduce speed by ~」が使用される．

> ✈ POINT:　速度調整が終了する場合
>
> 　以下の用語によって終了する.
>
> 1.「resume published speed」（公示された速度に従って下さい）
>
> 　SID, トランジション, STAR, 又は計器進入方式により飛行中の航空機（SID, トランジション, STAR, 若しくは, 計器進入方式との合流点に向かって通常航法により飛行中の航空機, すでに STAR を承認された航空機であって STAR の開始点より手前を通常航法により飛行中の航空機, 又はすでに計器進入方式を許可された航空機であって計器進入方式の開始点より手前を通常航法により飛行中の航空機を含む）を当該方式に公示された速度に従って飛行させる場合
>
> 2.「resume normal speed」（通常の速度に戻して下さい）
>
> 　上記以外の場合

　速度調整は,「resume published speed」若しくは「resume normal speed」の用語が通報される場合, 又は速度調整が自動的に終了する場合を除き, 次のセクター又は管制機関にハンドオフされた後も有効である.

PIL:	Tokyo Control, Nippon Air 715, maintain FL 290.
ACC:	Nippon Air 715, Tokyo Control, roger.
ACC:	Nippon Air 715, descend and maintain FL 270.
PIL:	Descend and maintain FL 270, Nippon Air 715.
ACC:	Nippon Air 715, descend to reach FL 170 by SAEKI, reduce speed to 250 knots.
PIL:	Descend to reach FL 170 by SAEKI, reduce speed to 250 knots, Nippon Air 715.
ACC:	Nippon Air 715, contact Kansai Radar 120.85.
PIL:	120.85, Nippon Air 715.

PIL: Kansai Radar, Nippon Air 715, leaving FL 190 descending FL 170, information Z.

RDR: Nippon Air 715, Kansai Radar, cleared via Alisa Bravo Arrival, descend and maintain 12,000.

PIL: Cleared via Alisa Bravo Arrival, descend 12,000, Nippon Air 715.

RDR: Nippon Air 715, and resume normal speed.

PIL: Resume normal speed, Nippon Air 715.

RDR: Nippon Air 715, contact Kansai Approach 120.25.

...MIHOU - Y-361 - SAEKI - Y-36 - ALISA

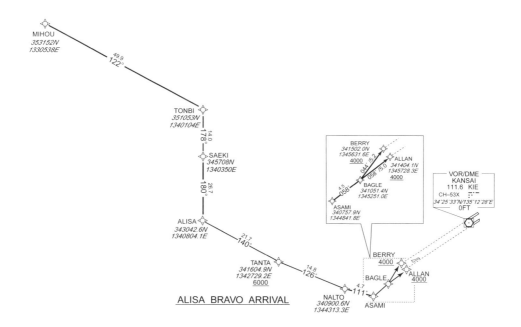

ALISA BRAVO ARRIVAL

✈ POINT: 速度調整が自動的に終了する場合

1．ホールディングが指示された場合

2．climb via SID to / descend via STAR to により，SID 若しくはトランジションによる上昇又は STAR による降下が指示された場合

3．進入許可が発出された場合

4．レーダー進入において接地点から5マイルの地点又は最終降下開始点のうちいずれか接地点から遠い方の地点を通過した場合

5．速度を維持すべき地点が明示されたのち当該地点を通過した場合

Phraseology Example 5

　QNH は気象機関が観測した値であり，管制機関・ATIS・広域対空援助局などから入手できる。ターミナル管制所から入手する場合は，以下のようになる．

PIL:　　Kansai Approach, Nippon Air 711, leaving FL 229 descend FL 160.

APP:　　Nippon Air 711, Kansai Approach, roger.

APP:　　Nippon Air 711, descend to reach 10,000 by DINAH, Osaka QNH 2988.

PIL:　　Descend to reach 10,000 by DINAH, QNH 2988, Nippon Air 711.

APP:　　Nippon Air 711, contact Kansai Approach 120.25.

PIL:　　Kumamoto Approach, Nippon Air 461, leaving FL 178 descending FL 150.

APP:　　Nippon Air 461, Kumamoto Approach, recleared direct IRPIN, descend and maintain 13,000, QNH 3031.

PIL:　　Recleared direct IRPIN, descend and maintain 13,000, Nippon Air 461.

APP:　　Nippon Air 461, request type of approach at Saga airport.

PIL:　　Expect RNP runway 29 approach via Irpin South Arrival, Nippon Air 461.

APP:　　Roger.

APP:　　Nippon Air 461, descend and maintain 7,000.

PIL:　　Descend and maintain 7,000, Nippon Air 461.

APP:　　Nippon Air 461, cleared via Irpin South Arrival, descend to reach 6,000 by IRPIN.

PIL:　　Cleared via Irpin South Arrival, descend to reach 6,000 by IRPIN, Nippon Air 461.

APP:　　Nippon Air 461, contact Fukuoka Radar 119.7.

PIL:　　Contact Fukuoka Radar 119.7, Nippon Air 461.

PIL:　　Fukuoka Radar, Nippon Air 461, leaving 12,100 desending 6,000.

RDR:　　Nippon Air 461, Fukuoka Radar, cleared for RNP runway 29 approach, cross IRPIN at 6,000, Saga QNH 3029.

管制区管制所からも入手することができる.

PIL:　　Contact Sapporo Control127.57, Nippon Air 155.

PIL:　　Sapporo Control, Nippon Air 155, FL 210.

ACC:　　Nippon Air 155, Sapporo Control, Aomori runway 06.

PIL:　　Nippon Air 155, request VOR Z runway 06 approach.

ACC:　　Nippon Air 155, roger, recleared direct YACHI via present position direct, stand by further clearance.

PIL:　　Direct YACHI, Nippon Air 155.

ACC:　　Nippon Air 155, descend at pilot's discretion maintain 11,000, area QNH 2994.

...KOHWA - Y-544 - SINGU - Y-542 - DATIS -RJOS

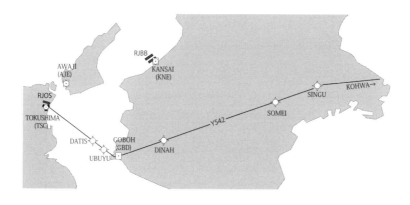

✈ POINT:　QNH　1

　　QNH を提供する管制機関の別により分類すると, 以下のようになる.

1. 管制区管制所　⇒　「area QNH」・「経路上の適切な地点の QNH」・「Baro-VNAV による進入を行う航空機に係る目的飛行場の QNH」

　　　　　　　　　　ただし, 広域セクターにおいて, 当該管轄区域内の飛行場への到着機に対しては「目的飛行場の QNH」

2. ターミナル管制所　⇒　「当該空域に関わる QNH」・「経路上の適切な地点の QNH」

　　　　　　　　　　ただし, 進入管制区内（セクター管轄区域内）の飛行場への到着機に対しては「目的飛行場の QNH」

3. 飛行場管制・着陸誘導管制所　⇒　当該機関の設置場所の QNH

Phraseology Example 6

管制上の都合等により待機の指示が発出される場合は，以下のようになる．

PIL:	Sapporo Control, Nippon Air 75, FL 390, we have 03Z weather at Kushiro.
ACC:	Nippon Air 75, Sapporo Control, Kushiro runway 17, request type of approach.
PIL:	Nippon Air 75, request ILS Z runway 17 approach.
ACC:	Nippon Air 75, roger, expect ILS Z runway 17 approach.
PIL:	Expect ILS Z runway 17 approach, Nippon Air 75.
ACC:	Nippon Air 75, recleared direct KANPO.
PIL:	Recleared direct KANPO, Nippon Air 75.
ACC:	Nippon Air 75, you are number 2, hold northwest of KANPO, expect approach at 0403.
PIL:	Hold northwest of KANPO, expect 0403, Nippon Air 75.
ACC:	Nippon Air 75, descend and maintain FL 210.
PIL:	Descend and maintain FL 210, Nippon Air 75.
ACC:	Nippon Air 75, descend and maintain FL 150.
PIL:	Descend and maintain FL 150, Nippon Air 75.
ACC:	Nippon Air 75, descend and maintain 8,000, area QNH 3026.
PIL:	3026, descend and maintain 8,000, Nippon Air 75.
PIL:	Sapporo Control, Nippon Air 75, start hold.
ACC:	Nippon Air 75, roger.
ACC:	Nippon Air 75, descend and maintain 7,000.

なお，航空機の運航上の都合等によって待機を希望する場合は，その旨を通報する．

PIL:　Tokyo Radar, Nippon Air 714, leaving 9,000 for 6,000, heading 310.

RDR:　Nippon Air 714, Tokyo Radar, roger.

RDR:　All stations, Tokyo Radar, runway 16R arrival, wind shear alert, 20 knots loss, 3 miles final, and 16L arrival, wind shear alert 20 knots loss, 3 miles final.

PIL:　Tokyo Radar, Nippon Air 714, request heading 280.

RDR:　Nippon Air 714, turn left heading 280, descend and maintain 3,000.

PIL:　Left heading 280, descend 3,000, Nippon Air 714.

PIL:　Tokyo Radar, Nippon Air 714, request heading 300 due to weather.

RDR:　Nippon Air 714, fly heading 300.

PIL:　Heading 300, Nippon Air 714.

PIL:　Tokyo Radar, Nippon Air 714, we want to delay the approach due to weather conditions, request stop descent 3,000, request holding north side of the airport.

RDR:　Nippon Air 714, roger, maintain 3,000, fly heading 360.

PIL:　Maintain 3,000, heading 360, Nippon Air 714.

RDR:　Nippon Air 714, expect hold over PIXUS, climb and maintain 4,000.

待機が予想される場合は，原則として待機させようとするフィックスの到着予定時刻の5分前までに待機指示が発出される．通常，以下の事項からなる．

1．待機経路の待機フィックスからの関係方位

2．待機フィックス（管制承認限界点と同一である場合は省略）

3．待機フィックスの入方向経路，若しくは入方向経路として使用する無線施設に関わる放射方位，コース，ベアリング，航空路又は経路

4．待機経路の出方向距離（DME使用の場合に限る）又は分を単位とする飛行時間

5．待機経路の旋回方向（右旋回の場合は省略）

ホールディングパターンが公示されている場合は，「hold northwest of KANPO」のように，上記3．～5．が省略される．

管制承認限界点への管制承認と同時に承認限界点での待機が指示されるときは，以下のように，待機フィックスが省略される．

「cleared to *** VOR, hold northeast」

ホールディングパターンが公示されていない場合（又は公示方法と異なる場合）は，

「hold north of COLOR on inbound track 196 degrees 1 minute left, left turns」

のようになる（旋回方向が特に指示されない場合は，右旋回を意味する）．

公示されている方法 　　　　　　　　公示方法と異なる場合

必要に応じて，進入予定時刻（EAT）や追加管制承認予定時刻（EFC）が提供される．なお，遅延時間が未定の場合は「delay not determined」が通報される．

APP:	Nippon Air 652, Tokyo Approach, runway 34L, cleared via Oshima 1A Arrival.
PIL:	Runway 34L, cleared via Oshima 1A Arrival, Nippon Air 652.
APP:	Nippon Air 652, hold west of Oshima, stand by EFC.
PIL:	Hold west of Oshima, Nippon Air 652.
APP:	Nippon Air 652, descend and maintain FL 180.
PIL:	Descend and maintain FL 180, Nippon Air 652.
APP:	Nippon Air 652, expect further clearance at 0945.
PIL:	EFC 0945, Nippon Air 652.
PIL:	Tokyo Approach, Nippon Air 652, start hold.

ホールディングを開始したとき，パイロットは自主的にその旨通報することが望ましい．なお，ホールディング中に降下のクリアランスを得た場合は，通常の降下率を維持して降下する．

待機が終了する場合としては,

1. 進入許可（ホールディング中にアプローチクリアランスを受領した場合は速やかに進入フィックスに向けて旋回し，最短の経路によって進入を開始する）

2. 待機フィックス以遠のフィックスへの直行指示

3. ヘディングの指示

4. 追加管制承認

5. 既承認経路での飛行指示（proceed via last routing cleared の指示）

の場合がある.

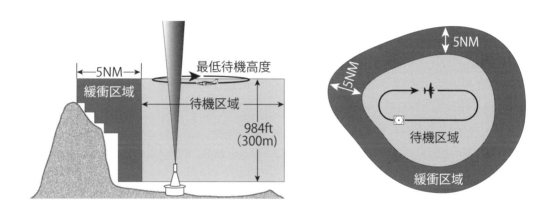

スタンダードパターンは右回りで 14,000 ft 以下の場合はアウトバウンド（フィックスを離れる方向）の飛行時間が 1 分（14,000 ft を超える場合は 1 分 30 秒）のものをいう. なお，公示されたホールディングパターンには MHA（Minimum Holding Altitude）が設定されており，待機区域内の地上障害物から最小 984 ft の垂直間隔を確保し，待機区域の周辺 5 マイルの緩衝区域内の障害物も考慮されている.

ホールディングエントリーは 3 つの方式があり，ホールディングフィックスに到達したときのヘディングにより使い分ける. もしヘディングが象限の境目付近であるときは，それが 5° 以内ならどちらの方式を使用してもよい.

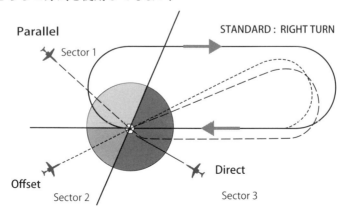

Phraseology Example 7

進入フィックス以遠への飛行について遅延が予想されない場合であって，進入フィックス到達の５分前までに進入許可が発出されない場合は，必要に応じ「no delay expected」の用語により遅延がない旨が通報される．なお，当該機が進入フィックスへ到着するまでに進入許可が発出される．

PIL:	Sapporo Control, Nippon Air 75, FL 370, we have 2300Z Kushiro weather.
ACC:	Nippon Air 75, Sapporo Control, Kushiro runway 17, request type of approach.
PIL:	Request RNP Z runway 17 approach, Nippon Air 75.
ACC:	Nippon Air 75, roger, expect RNP Z runway 17 approach.
PIL:	Nippon Air 75.
PIL:	Sapporo Control, Nippon Air 75, request descent.
ACC:	Nippon Air 75, descend and maintain FL 210.
PIL:	Descend and maintain FL 210, Nippon Air 75.
ACC:	Nippon Air 75, recleared direct CRANE, request CRANE estimate.
PIL:	Recleared direct CRANE, estimate CRANE 0403, Nippon Air 75.
ACC:	Nippon Air 75, roger.
ACC:	Nippon Air 75, descend and maintain FL 150.
PIL:	Descend and maintain FL 150, Nippon Air 75.
ACC:	Nippon Air 75, descend and maintain 9,000, area QNH 2964, no delay expected.
PIL:	Descend and maintain 9,000, 2964, Nippon Air 75.
ACC:	Nippon Air 75, descend and maintain 3,600, maintain 3,600 until CRANE, cleared for RNP Z runway 17 approach.
PIL:	Descend and maintain 3,600, maintain 3,600 until CRANE, cleared for RNP Z runway 17 approach, Nippon Air 75.
ACC:	Nippon Air 75, Kushiro QNH 2964, contact Kushiro Tower 118.05.

（参考1）クルーズ飛行方式によるクリアランスが発出される場合は，以下のように，管制承認発出時等に「maintain」の代わりに「cruise」の用語が使用される場合がある．このクリアランスには巡航に続く進入許可を含んでいる．

MEA 以上の高度である限りは，承認された高度以下の任意の高度を選ぶことができる．パイロットの判断で随時降下して，所定の進入フィックス到達後は計器進入方式を開始してもよい．

なお，いったん離脱高度を通報した場合は特に指示のない限り通過した高度に戻ってはならない．主に，離島に向かう航空路等で使用される場合が多い．

PIL: Hakodate Tower, Nippon Air 895, with information E, spot 6.

TWR: Nippon Air 895, Hakodate Tower, go ahead.

PIL: Nippon Air 895, request clearance to Okushiri airport, request altitude change 12,000.

TWR: Nippon Air 895, roger, propose altitude 12,000, stand by clearance.

PIL: Stand by, Nippon Air 895.

TWR: Nippon Air 895, clearance.

PIL: Nippon Air 895, go ahead.

TWR: Nippon Air 895 cleared to Okushiri airport via Okushiri One Departure, flight planned route, cruise 12,000, Radar frequency 119.0, squawk 4623.

PIL: Nippon Air 895 cleared to Okushiri airport via Okushiri One Departure, flight planned route, cruise 12,000, Radar frequency 119.0, squawk 4623.

TWR: Nippon Air 895, read back is correct.

PIL: Hakodate Tower, Nippon Air 895, request taxi.

TWR: Nippon Air 895, one eighty turn approved, you are number 2.

PIL: One eighty turn approved, Nippon Air 895.

TWR: Nippon Air 895, taxi to holding point runway 12, follow ANA Boeing 737, you are number 2 departure.

PIL: Nippon Air 895, taxi to holding point runway 12, follow Boeing 737.

PIL:　　Nippon Air 895, ready.

TWR:　　Nippon Air 895, do you accept T-2 intersection.

PIL:　　Affirm, Nippon Air 895.

TWR:　　Nippon Air 895, roger, runway 12 at T-2, line up and wait, inbound traffic 9 miles on final.

PIL:　　Runway 12 at T-2, line up and wait, Nippon Air 895.

TWR:　　Nippon Air 895, wind 160 at 11, runway 12 cleared for take-off.

PIL:　　Runway 12 cleared for take-off, Nippon Air 895.

TWR:　　Nippon Air 895, contact Radar.

PIL:　　Contact Radar, Nippon Air 895.

PIL:　　Hakodate Radar, Nippon Air 895, leaving 1,800 climbing 12,000.

RDR:　　Nippon Air 895, Hakodate Radar, radar contact.

RDR:　　Nippon Air 895, contact Okushiri Radio 122.7.

PIL:　　Okushiri Radio 122.7, Nippon Air 895.

PIL:　　Okushiri Radio, Nippon Air 895.

AFIS:　　Nippon Air 895, Okushiri Radio, go ahead.

PIL:　　Nippon Air 895, cruise 12,000, make VOR A approach straight in landing runway 31, estimate MAIKA 47.

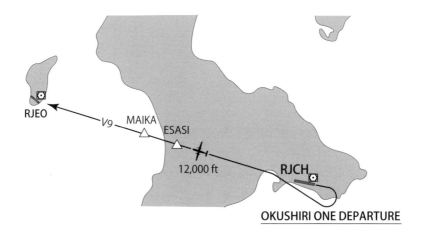

UNIT.5. Descent & Approach

Words & Phrases

vector to ~	vector to final approach course
～への誘導を行います	最終進入コースへの誘導を行います
expect ~ approach	cleared for approach
～進入を予期して下さい	進入を許可します
report high (low) station	
ハイ（ロー）ステーション離脱を通報して下さい	
report starting base (procedure) turn	report completing base (procedure) turn
基礎（方式）旋回の開始を通報して下さい	基礎（方式）旋回の終了を通報して下さい
circle to runway ~	cleared for contact approach at or below ~
滑走路～へ周回進入を行って下さい	～以下で目視進入を許可します
cleared for contact approach (if not possible maintain ~ and advise)	
目視進入を許可します（不可能の場合は～して通報して下さい）	
cleared visual approach runway ~	reduce to minimum approach speed
滑走路～への視認進入を許可します	最低進入速度に減速して下さい
reduce to minimum clean speed (*1)	
ミニマムクリーンスピードに減速して下さい	

† （*1）ミニマムクリーンスピードは，高揚力装置，スピードブレーキ及び着陸装置を展開することなく飛行可能な速度であり，ターボジェット機の場合は FL 150 未満において通常 220 ノット前後である.

Introduction

IFR による到着は，一般的に STAR 又はレーダー誘導に続く計器進入によって行われる．

STAR（標準計器到着方式）とは，IFR で到着する航空機が ATS 経路から計器進入方式等を開始するまでを定めた飛行経路，旋回方向，高度等の方式であり，計器進入方式は IFR で到着する航空機が飛行する経路と高度等の一連の方式である．

計器進入は，通常，到着経路・初期進入・中間進入・最終進入・進入復行の 5 つの部分に分けられる．それぞれの飛行区分は「セグメント」と呼ばれ，実際にはすべてのセグメントが設定されているわけではなく，STAR を経て最終進入につながる方式と，基礎旋回を経て最終進入につながる方式の 2 つが多い．なお，各空港には進入してくる方向ごとに STAR が設定されている場合が多い．

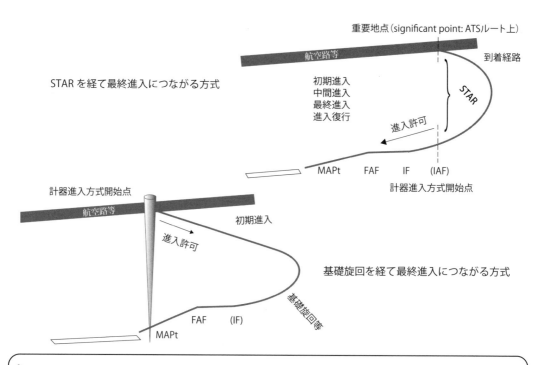

POINT:　IFR 到着機の動き－1

　IFR での到着機は，飛行場においてどのような管制業務が実施されているかにより，進入方法に違いがある．（以下，主なもの）

1．航空路等　⇒　STAR　⇒　計器進入方式　⇒　滑走路
2．航空路等　⇒　無線施設（VOR 等）上空　⇒　計器進入方式　⇒　滑走路
3．航空路等　⇒　計器進入方式　⇒　滑走路
4．航空路等　⇒　レーダー誘導　⇒　計器進入方式　⇒　滑走路
5．航空路等　⇒　レーダー誘導　⇒　視認進入　⇒　滑走路
6．航空路等　⇒　レーダー誘導　⇒　経路指定視認進入　⇒　滑走路
7．航空路等　⇒　目視進入　⇒　滑走路

POINT: 進入管制業務

進入管制業務は ACC が行う場合と，空港の進入管制所で行う 2 通りがある．

1．ACC が行う進入管制業務

- 広域レーダー進入管制業務　⇒　北海道東部（女満別・中標津・釧路・帯広）や東北地方等（青森・大館能代・秋田）で進入フィックスまで ACC のレーダーで行う進入管制業務→下図の A-1
- それ以外　⇒　タワーのみの飛行場管制業務が行われている空港，レディオ又はリモート等に出入域する IFR 機に対する進入管制業務→下図の A-2

2．ターミナル管制所が行う進入管制業務

- ターミナルレーダー管制業務
- レーダーを使用しない管制業務（主にレーダー停波時等）

A－1

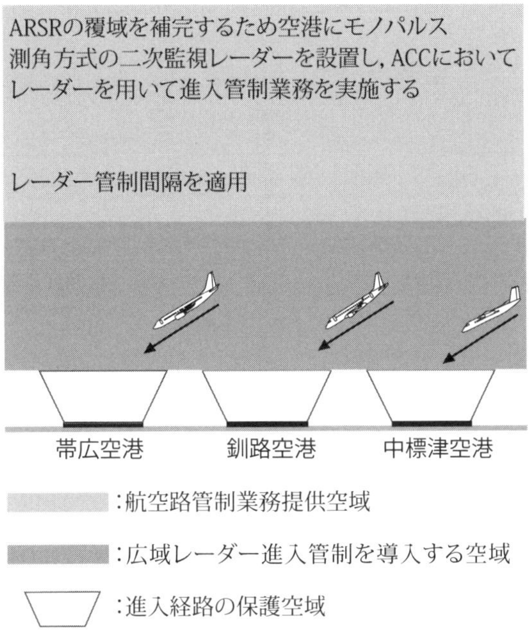
ARSRの覆域を補完するため空港にモノパルス
測角方式の二次監視レーダーを設置し, ACCにおいて
レーダーを用いて進入管制業務を実施する
レーダー管制間隔を適用
帯広空港
釧路空港
中標津空港
:航空路管制業務提供空域
:広域レーダー進入管制を導入する空域
:進入経路の保護空域

レーダー管制間隔があるので，
それぞれの空港へ進入可能

原則，最終進入コースへの誘導や
視認進入は行われない

A－2

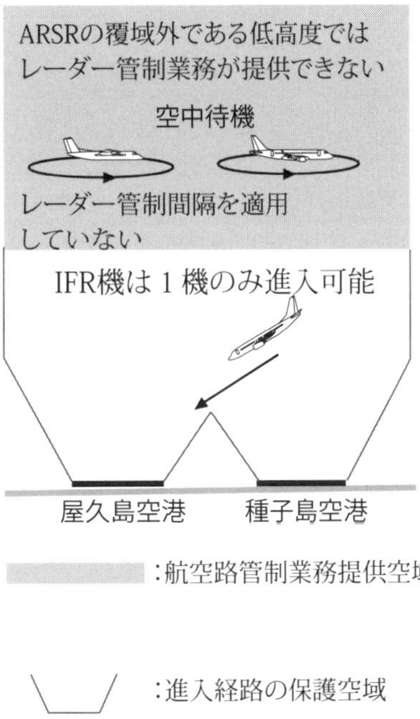
ARSRの覆域外である低高度では
レーダー管制業務が提供できない
空中待機
レーダー管制間隔を適用
していない
IFR機は 1 機のみ進入可能
屋久島空港
種子島空港
:航空路管制業務提供空域
:進入経路の保護空域

出発機は到着機が着陸するまで離陸できない．
また，到着機も先行到着機が着陸するまで進
入をできないため，上空待機

Basic Example

ターミナル管制所入域管制官（Kansai Approach）と交信

> **PIL:** Kansai Approach, Nippon Air 710, leaving FL 154 descending 13,000, information I.
>
> **APP:** Nippon Air 710, Kansai Approach, fly heading 250 vector to IKOMA, descend and maintain 8,000, reduce speed to 230 knots.
>
> **PIL:** Fly heading 250, descend 8,000, reduce 230, Nippon Air 710.
>
> **APP:** Nippon Air 710, turn right heading 270, descend and maintain 6,000.
>
> **PIL:** Right heading 270, descend 6,000, Nippon Air 710.
>
> **APP:** Nippon Air 710, reduce speed to 210 knots.
>
> **PIL:** Reduce speed to 210 knots, Nippon Air 710.
>
> **APP:** Nippon Air 710, resume own navigation direct IKOMA.
>
> **PIL:** Resume own navigation direct IKOMA, Nippon Air 710.
>
> **APP:** Nippon Air 710, descend and maintain 5,000.
>
> **PIL:** Descend 5,000, Nippon Air 710.
>
> **APP:** Nippon Air 710, reduce speed to 190 knots, descend and maintain 4,000.
>
> **PIL:** Speed to 190 knots, descend 4,000, Nippon Air 710.
>
> **APP:** Nippon Air 710, descend and maintain 3,500, 7 miles MIDOH, cleared for ILS runway 32L approach.
>
> **PIL:** Descend 3,500, cleared for ILS runway 32L approach, Nippon Air 710.
>
> **APP:** Nippon Air 710, contact Osaka Tower 118.1.
>
> **PIL:** Contact Osaka Tower 118.1, Nippon Air 710.

STANDARD ARRIVAL CHART-INSTRUMENT

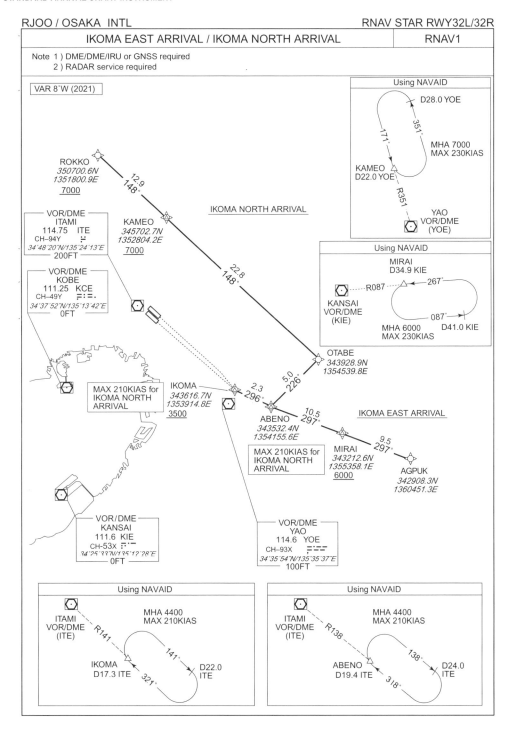

Phraseology Example 1

　通常，進入管制区の境界に近づいた時点で，パイロットはアプローチコントロールへ周波数の切り替えを指示される．アプローチコントロールとのイニシャルコンタクトは，高度変更中であれば承認高度と通過高度を，水平飛行中であれば維持している高度をそれぞれ100 ft単位で（ATISが運用されている場合，コードも合わせて）通報する．

　STARの承認と進入許可が同時に発出される場合は，特に必要な場合を除き高度は指示されず，パイロットはATS経路の最低経路高度（MEA）及びSTARの高度制限又は速度に従って降下し進入を行う．

　なお，航空路，RNAV5経路及び直行経路を航行中の場合，進入フィックス上空到達以前に降下の指示を含まない進入許可が発出された場合は，パイロットは航空路，RNAV5経路及び直行経路の最低経路高度まで降下することができる．

　進入許可は，ターミナル管制所のある飛行場ではアプローチから，その他の飛行場ではACC又はタワー（レディオ，リモートも含む）から発出される．

ACC:　Nippon Air 619, descend and maintain 12,000.

PIL:　Descend and maintain 12,000, Nippon Air 619.

ACC:　Nippon Air 619, contact Kagoshima Radar 121.4.

PIL:　Kagoshima Radar 121.4, Nippon Air 619.

PIL:　Kagoshima Radar, Nippon Air 619, maintain 12,000, information T.

RDR:　Nippon Air 619, Kagoshima Radar, cleared for ILS Z runway 27 approach via Ryugu Arrival.

PIL:　Cleared for ILS Z runway 27 approach via Ryugu Arrival, Nippon Air 619.

RDR:　Nippon Air 619, contact Miyazaki Tower 118.3.

PIL:　Contact Miyazaki Tower 118.3, Nippon Air 619.

PIL:　Miyazaki Tower, Nippon Air 619, 2 miles to OYODO.

TWR:　Nippon Air 619, Miyazaki Tower, runway 27 cleared to land, wind 250 at 13.

　交通量の多い所では，通常，複数の計器進入方式のうちの1つが管制機関から指定される．この場合，「cleared for ILS Z runway 27 approach」のような進入許可が発出される．異なる進入方式を希望する場合は，別途要求しなければならない．

STANDARD ARRIVAL CHART - INSTRUMENT

RJFM / MIYAZAKI   RNAV STAR

RYUGU ARRIVAL	RNAV1

Note 1) DME/DME/IRU or GNSS required.
     2) RADAR service required.

VAR 7˚W (2020)

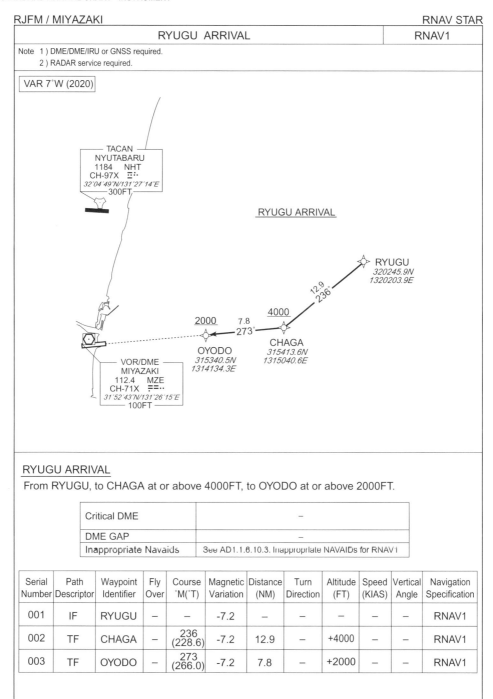

RYUGU ARRIVAL

From RYUGU, to CHAGA at or above 4000FT, to OYODO at or above 2000FT.

Critical DME	–
DME GAP	–
Inappropriate Navaids	See AD1.1.6.10.3. Inappropriate NAVAIDs for RNAV1

Serial Number	Path Descriptor	Waypoint Identifier	Fly Over	Course ˚M(˚T)	Magnetic Variation	Distance (NM)	Turn Direction	Altitude (FT)	Speed (KIAS)	Vertical Angle	Navigation Specification
001	IF	RYUGU	–	–	-7.2	–	–	–	–	–	RNAV1
002	TF	CHAGA	–	236 (228.6)	-7.2	12.9	–	+4000	–	–	RNAV1
003	TF	OYODO	–	273 (266.0)	-7.2	7.8	–	+2000	–	–	RNAV1

Phraseology Example 2

　飛行経路について STAR が承認された場合は，進入フィックスまでの経路とともに当該 STAR に公示された高度制限が追加されたことを意味する．下記のように，STAR の承認に進入許可を伴わない場合は，高度について，必要に応じ追加や変更等が指示される．

RDR:　　Nippon Air 619, Kagoshima Radar, cleared via Ryugu Arrival, descend and maintain 10,000.

PIL:　　Cleared via Ryugu Arrival, descend and maintain 10,000, Nippon Air 619.

RDR:　　Nippon Air 619, descend via STAR to 2,000, cleared for ILS Z runway 27 approach.

PIL:　　Descend via STAR to 2,000, cleared for ILS Z runway 27 approach, Nippon Air 619.

RDR:　　Nippon Air 619, contact Miyazaki Tower 118.3.

　STAR の高度制限又は速度に従って降下する必要がある場合は，「descend via STAR to ~」（STAR の制限に従い~まで降下して下さい）の用語が使用される．この場合，降下の時機はパイロットに任せられる．なお，それまで行われていた速度調整は自動的に終了する．

　✈ POINT:　IFR 到着機の動き−2

　　進入を開始するまでの飛行方法には，以下のものがある．

1．計器進入を開始するフィックスへのレーダー誘導
2．STAR による経路の指定
3．計器進入を開始するフィックスへの直行指示
4．最終進入コースへのレーダー誘導
5．視認進入のためのレーダー誘導

　上記3．のみ「予定される計器進入方式」の確認が必要である．また，計器進入を開始するフィックスへの経路が明確でない場合は，「管制機関から計器進入方式が通報された場合」，又は，「ATIS に含まれている場合（パイロットが通報する）」には，当該進入方式を開始するフィックス等が承認経路となる．

INSTRUMENT APPROACH CHART

RJFM / MIYAZAKI   ILS Z or LOC Z RWY27

KAGOSHIMA APP 121.4 – 362.3 120.9 – 261.2	ILS-LOC 108.9 IMZ ≡=.. ILS-GP 329.3 ILS-DME IMZ CH-26X	MIYAZAKI TOWER 118.3-126.2 123.6-261.2	RADAR AVBL ATIS 126.8

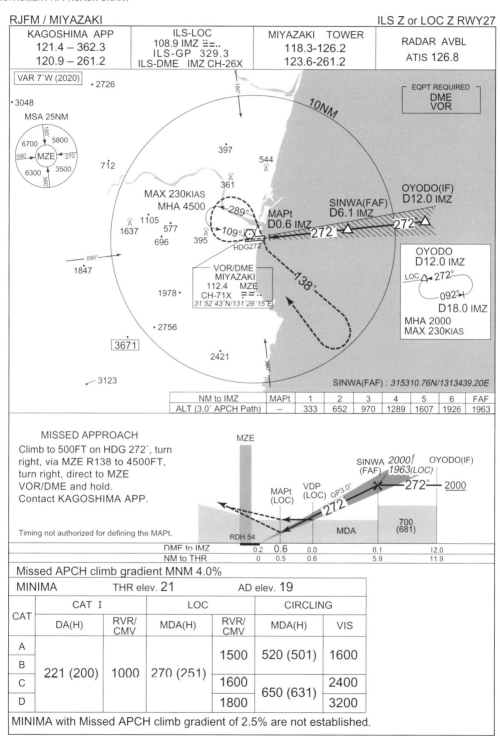

SINWA(FAF) : 315310.76N/1313439.20E

NM to IMZ	MAPt	1	2	3	4	5	6	FAF
ALT (3.0° APCH Path)	–	333	652	970	1289	1607	1926	1963

MISSED APPROACH

Climb to 500FT on HDG 272°, turn
right, via MZE R138 to 4500FT,
turn right, direct to MZE
VOR/DME and hold.
Contact KAGOSHIMA APP.

Timing not authorized for defining the MAPt.

DME to IMZ	0.2	0.6	0.0	0.1	12.0
NM to THR	0	0.5	0.6	5.9	11.9

Missed APCH climb gradient MNM 4.0%

MINIMA   THR elev. 21   AD elev. 19

CAT	CAT I		LOC		CIRCLING	
	DA(H)	RVR/ CMV	MDA(H)	RVR/ CMV	MDA(H)	VIS
A	221 (200)	1000	270 (251)	1500	520 (501)	1600
B						
C				1600	650 (631)	2400
D				1800		3200

MINIMA with Missed APCH climb gradient of 2.5% are not established.

Phraseology Example 3

STAR の経路によらず，レーダー誘導，又は，経路の変更等によって進入を行う場合もある．進入フィックスへ直行する指示で，その後，進入許可が発出される場合は，以下のようになる．

PIL:　　Kagoshima Radar, Nippon Air 619, leaving 13,800 descending 12,000, information E, request RNP Z runway 09 approach.

RDR:　　Nippon Air 619, Kagoshima Radar, roger, recleared direct HIMKA, descend and maintain 3,500.

PIL:　　Recleared direct HIMKA, descend and maintain 3,500, Nippon Air 619.

RDR:　　Nippon Air 619, cleared for RNP Z runway 09 approach, cross HIMKA at or above 3,500.

PIL:　　Cleared for RNP Z runway 09 approach, cross HIMKA at or above 3,500, Nippon Air 619.

RDR:　　Nippon Air 619, contact Miyazaki Tower 118.3.

PIL:　　Contact Miyazaki Tower 118.3, Nippon Air 619.

PIL:　　Miyazaki Tower, Nippon Air 619, approaching HIMKA, RNP Z runway 09.

TWR:　　Nippon Air 619, Miyazaki Tower, runway 09 cleared to land, wind 140 at 9.

上記のように，進入フィックスへ直行する指示で，進入許可が発出される場合は，多くの場合，直行フィックスに到達するまでに進入許可が発出される．

RNP 進入及び RNP AR 進入の場合，誘導を終了する場合（resume own navigation direct ~）や管制承認をフィックスを経由するものに変更する場合（recleared direct ~）は，進入許可発出時にフィックスまで維持すべき高度が，「cross ~ at or above ~」「maintain ~ until ~」「descend and maintain ~」「maintain ~」等の用語により指示される．

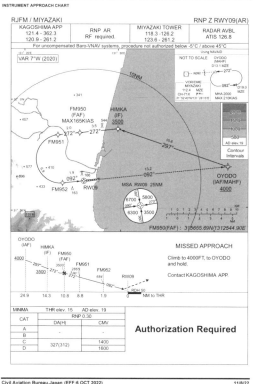

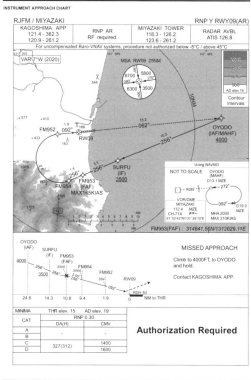

	レーダー誘導	直行	レーダー誘導
1	passing FL 156 descend 12,000, information E, request RNP Y runway 09 approach fly heading 240 vector to SURFU, descend and maintain 3,500	leaving 13,800 descending 12,000, information T recleared direct OYODO, descend and maintain 3,000	leaving 13,800 descending 12,000, information T fly heading 210 vector to OYODO for spacing, descend and maintain 10,000
2			descend and maintain 6,000
3			turn right heading 250
4	resume own navigation direct SURFU		descend and maintain 3,000
5			descend and maintain 2,000
6			resume own navigation direct OYODO
7	cleared for RNP Y runway 09 approach, cross SURFU at or above 3,500	20 miles from OYODO, descend and maintain 2,000, cleared for ILS Z runway 27 approach, cross OYODO at or above 2,000	10 miles from OYODO, clcared for ILS Z runway 27 approach, cross OYODO at or above 2,000

61

Phraseology Example 4

　最終進入コースにレーダー誘導される場合は，以下のようになる．このような，計器進入方式の最終進入コースへの誘導は「vector to final」と呼ばれる．なお，誘導によって飛行経路が延長される場合には通常「for spacing」（＝間隔設定のために）の用語が使用される．

RDR:　Nippon Air 619, Kagoshima Radar, fly heading 220 vector to final approach course, descend and maintain 3,000.

PIL:　Nippon Air 619, fly heading 220, descend and maintain 3,000.

RDR:　Nippon Air 619, reduce speed to 210 knots.

PIL:　Nippon Air 619, reduce speed to 210 knots.

RDR:　Nippon Air 619, turn left heading 210.

PIL:　Nippon Air 619, turn left heading 210.

RDR:　Nippon Air 619, expect vector across final approach course for spacing.

PIL:　Nippon Air 619.

RDR:　Nippon Air 619, turn right heading 300.

PIL:　Nippon Air 619, turn right heading 300.

RDR:　Nippon Air 619, descend and maintain 2,000.

PIL:　Nippon Air 619, descend and maintain 2,000.

RDR:　Nippon Air 619, reduce speed to 190 knots.

PIL:　Nippon Air 619, reduce speed to 190 knots.

RDR:　Nippon Air 619, turn left heading 240.

PIL:　Nippon Air 619, turn left heading 240.

RDR:　Nippon Air 619, 2 miles from SINWA, cleared for ILS Z runway 27 approach, contact Miyazaki Tower 118.3.

PIL: Nippon Air 619, cleared for ILS Z runway 27 approach, contact Miyazaki
        Tower 118.3.

PIL: Miyazaki Tower, Nippon Air 619, approaching SINWA.

　最終進入コースへの誘導の場合，スムーズに実施するため，最終進入フィックスの手前で，オーバーシュートすることがないよう，比較的浅い角度でコースへ会合するよう誘導が行われる．

　最終進入コースへの会合はアプローチゲート以遠で行われる．アプローチゲートとは，最終進入コース上において，

　　　滑走路進入端から5マイルの点

又は

　　　最終進入フィックスから飛行場の反対側へ1マイル

のいずれかのうち，滑走路から遠い方をいう．

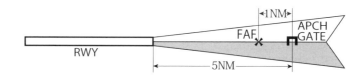

　特に飛行場の雲高の値に飛行場標高を加えた高さが MVA+500ft 未満，或いは，地上視程が5km未満のときは，アプローチゲートより2マイル以遠の地点で会合される．

　最大会合角は，会合地点がアプローチゲートから2マイル未満の場合は20°，2マイル以遠の場合は30°となっている．

　最終進入コースへ誘導されて進入許可が発出された場合は，当該進入方式に初期進入（基礎旋回／方式旋回）が設定されていても，それらのセグメントは省略し最終進入のみを行う．

✈ POINT: レーダー誘導の終了

　レーダー誘導は，通常，以下の2つの場合に終了する．

1．「resume own navigation（direct）」の用語
　　・必要に応じて，位置と飛行方法が指示される
2．進入許可の発出（誘導の終了「resume own navigation」は通報されない）
　　・視認進入（cleared visual approach）
　　・計器進入方式の最終進入コース（cleared for ～ approach）

Phraseology Example 5

視認進入の場合は，以下のようになる．

RDR: Nippon Air 619, Kagoshima Radar, fly heading 240 vector to right downwind runway 09 for visual approach, descend and maintain 6,000.

PIL: Fly heading 240, descend and maintain 6,000, Nippon Air 619.

RDR: Nippon Air 619, descend and maintain 2,000.

PIL: Descend 2,000, Nippon Air 619.

RDR: Nippon Air 619, turn right heading 250, descend and maintain 1,600.

PIL: Right heading 250, descend and maintain 1,600, Nippon Air 619.

RDR: Nippon Air 619, turn right heading 270, report airport in sight.

PIL: Right heading 270, report airport in sight, Nippon Air 619.

RDR: Nippon Air 619, airport 2 o'clock, 8 miles, report airport in sight.

PIL: Nippon Air 619, airport in sight.

RDR: Nippon Air 619, cleared visual approach runway 09, join right downwind, contact Miyazaki Tower 118.3.

✈ POINT:　視認進入 (visual approach) と 目視進入 (contact approach)

　天候が良好なときは，IFR を維持したまま目視により所定の計器進入方式によらないで効率のよい進入を行うことができる．このうち，主に運用されているのが，視認進入と目視進入である．それぞれ，実施状況・環境が異なっている．

1．視認進入

- ・レーダー管制
- ・雲高の値に飛行場標高を加えた高さが最低誘導高度よりも 500 ft 以上高く，かつ，地上視程が 5 km 以上あること
- ・ATC の判断（パイロットの要求可）

2．目視進入

- ・ノンレーダー
- ・雲高の値に飛行場標高を加えた高さが進入開始高度よりも高く，かつ，地上視程が 1,500 m 以上であること
- ・パイロットの要求

視認進入；

IFR での目視による進入で，飛行場（の場周経路）へのレーダー誘導が行われ，飛行場に近づいた後に行う進入

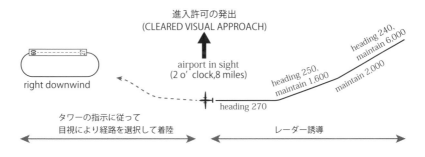

　視認進入のために誘導が行われる場合は，「〜 vector to（runway 09）traffic pattern」「〜 vector for visual approach runway 09」等の用語が使用される．

　通常，進入許可は以下の要領で発出される．

１．先行機がある場合：先行機の交通情報の後，

　１−１．先行機を視認 ⇒ 先行機に続く（follow the traffic）⇒（位置情報）

　１−２．先行機を視認できない ⇒ 飛行場視認の通報 ⇒ 先行機の交通情報 ⇒（位置情報）

の後，進入許可が発出される．なお，

　　A．先行機が着陸

　　B．当該機が先行機を視認する

　　C．飛行場（ローカル）管制席（官）が当該機を視認する

まではアプローチの周波数により指示を受ける．

２．先行機がない場合

　２−１．飛行場視認の通報 ⇒（位置情報）⇒ 進入許可

　進入許可（cleared visual approach）発出後，視認進入（又は経路指定視認進入）においては，

　　１．VMC の維持　　　　　　　　　　２．地上障害物との衝突防止

　　３．視認している関連機との間隔維持　　４．後方乱気流回避

等は，パイロットの責任において行わなければならない．

　なお，目視進入（P.70 参照）においては，他の IFR 機との衝突防止は管制間隔によって確保されているので，

　　１．VFR で飛行中の他の航空機との間隔維持

　　２．障害物との衝突防止

をパイロットの責任において行わなければならない．

Phraseology Example 6

　飛行場の地形，障害物，或いは滑走路と航空保安無線施設との関係位置等によって，特定滑走路へのストレートインランディングによる進入方式が設定されていない場合は周回進入によって当該滑走路への着陸が行われる．特定の進入方式に続く周回が指示される場合は，進入方式名に周回後着陸する滑走路名がつけられる．

RDR: Nippon Air 619, Kagoshima Radar, cleared via Ryugu Arrival, expect circling approach runway 09.

PIL: Nippon Air 619, cleared via Ryugu Arrival.

RDR: Nippon Air 619, descend and maintain 10,000.

PIL: Nippon Air 619, descend and maintain 10,000.

RDR: Nippon Air 619, descend and maintain 6,000.

PIL: Nippon Air 619, descend and maintain 6,000.

RDR: Nippon Air 619, descend via STAR to 2,000, cleared for ILS Z runway 27 approach, circle to runway 09.

PIL: Nippon Air 619, descend via STAR to 2,000, cleared for ILS Z runway 27 approach, circle to runway 09.

RDR: Nippon Air 619, contact Miyazaki Tower 118.3.

PIL: Nippon Air 619, contact Miyazaki Tower 118.3.

PIL: Miyazaki Tower, Nippon Air 619, departed OYODO, runway 09.

TWR: Nippon Air 619, Miyazaki Tower, report left break, wind 150 at 10.

PIL: Nippon Air 619, report left break.

PIL: Nippon Air 619, left break.

TWR: Nippon Air 619, runway 09 cleared to land, wind 150 at 11.

　パイロットは，飛行場又は滑走路等を視認した後，必要であれば，飛行場視認又は滑走路視認の通報をする．適切な地点より最終進入経路を離脱し，周回進入区域（circling area）内を飛行するが，周回進入区域を逸脱しないようにしなければならない．

　また，管制機関から飛行場又は滑走路等の視認の通報を求められる場合もある．

PIL:　　Miyazaki Tower, Nippon Air 619, over OYODO, circling 09.

TWR:　　Nippon Air 619, Miyazaki Tower, runway 09, report runway in sight.

PIL:　　Nippon Air 619, report runway in sight.

PIL:　　Nippon Air 619, runway in sight.

TWR:　　Nippon Air 619, report left break.

PIL:　　Nippon Air 619, report left break.

PIL:　　Nippon Air 619, left break.

　通常，最終進入中に飛行場又は滑走路等を視認した後，目視により周回進入区域内に飛行経路を選定するが（周回進入区域は，航空機の進入カテゴリー によって異なる），天候が良い場合等は，進入の早い段階から通常の計器進入の最終進入コースを外れ，場周経路を目指して飛行することが多い．

　また，周回進入の最低気象条件（circling minima）に近い場合は，航空機は飛行場を飛行視程内に保つように周回進入区域内に経路を選定して飛行する（ミニマムサークリング）．

　周回中，着陸のための目視による降下を開始するまで適用される最低高度として MDA が設定されている．

　周回進入における進入の限界点及び MAPt は，MDA 降下後，滑走路末端から当該周回進入に関わる最低気象条件の飛行視程に相当する距離の点であり，周回進入に適用される最低気象条件はストレートインミニマよりも更に一段高い値の地上視程が適用され，RVR は適用されない．

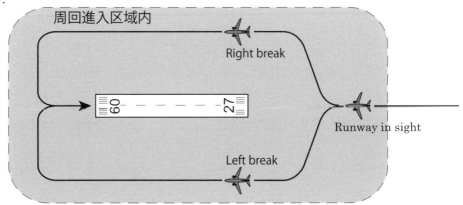

67

Phraseology Example 7

　進入許可は，交通状況により，公示されている計器進入方式が指定される場合（「cleared for ~ approach」）と，計器進入方式をパイロットが選択できる場合（「cleared for approacch」）とがある．RNP AR 進入方式が設定されていない飛行場対空援助業務が実施される空港等では，ACC が計器進入方式の種類を指定しない進入許可「cleared for approach」を発出する場合が多い．

ACC: 　Nippon Air 81, cleared via Okuji Arrival, descend and maintain 9,000.

PIL: 　Cleared via Okuji Arrival, descend and maintain 9,000, Nippon Air 81.

ACC: 　Nippon Air 81, cleared for approach to Fukushima airport, contact Fukushima Radio 118.05.

PIL: 　Cleared for approach to Fukushima airport, contact Fukushima Radio 118.05, Nippon Air 81.

PIL: 　Fukushima Radio, Nippon Air 81.

AFIS: 　Nippon Air 81, Fukushima Radio, go ahead.

PIL: 　Nippon Air 81, we have approach clearance, we will make ILS Y runway 01 approach via Okuji Arrival, estimate SOUMA 40.

AFIS: 　Nippon Air 81, using runway 01, wind 040 degrees at 3 knots, temperature 5, QNH 3022, traffic not reported, report SOUMA.

PIL: 　QNH 3022, report SOUMA, Nippon Air 81.

AFIS: 　Nippon Air 81, ATC requests present altitude.

PIL: 　4,800, Nippon Air 81.

AFIS: 　Nippon Air 81, roger.

PIL: 　Fukushima Radio, Nippon Air 81, departed SOUMA.

　進入許可の発出においては通常，すべての進入方式が許可されるわけではなく，クリアランスリミットとして指定されたフィックス，又は，飛行計画で明示した経路上のフィックスを経由する進入方式が，トラフィックや気象状況に応じて認められる場合が多い．

　なお，進入開始点までの経路に関しては MEA を遵守しなければならない．

Phraseology Example 8

RNP AR 進入方式が設定されている飛行場では，管制機関がアプローチタイプを確認・指定して，公示されている計器進入方式を指定して進入許可が発出される.

PIL:	Tokyo Control, Nippon Air 291, FL 340.
ACC:	Nippon Air 291, Tokyo Control, descend and maintain FL 240.
PIL:	Descend and maintain FL 240, Nippon Air 291.
ACC:	Nippon Air 291, request your approach type.
PIL:	Nippon Air 291, ILS runway 10 approach.
ACC:	Nippon Air 291, roger, descend and maintain 7,000, area QNH 2960.
PIL:	Descend and maintain 7,000, 2960, Nippon Air 291.
ACC:	Nippon Air 291, maintain 7,000 until passing Tottori, then cleared for ILS runway 10 approach.
PIL:	Maintain 7,000 until passing Tottori, then cleared for ILS runway 10 approach, Nippon Air 291.
ACC:	Nippon Air 291, contact Tottori Radio 118.15.
PIL:	Tottori Radio 118.15, Nippon Air 291.
PIL:	Tottori Radio, Nippon Air 291.
AFIS:	Nippon Air 291, Tottori Radio, go ahead.
PIL:	Nippon Air 291, we have approach clearance, we will make ILS runway 10 approach, estimate over Tottori 35, we have 0300 Tottori weather.
AFIS:	Nippon Air 291, using runway 10, wind 180 degrees at 3 knots, crosswind, temperature 24, QNH 2960 inches, traffic not reported around Tottori airport, report high station.
PIL:	Report over high station, 2960, Nippon Air 291.
PIL:	Tottori Radio, Nippon Air 291, leaving high station.
AFIS:	Nippon Air 291, wind 170 degrees at 3 knots, report base turn inbound.

Phraseology Example 9

　飛行時間を短縮するため，又は，IFR 出発機がある場合等，航空交通を考慮した効率よい流れを作るため，IFR をキャンセルする場合がある．この場合，IFR をキャンセルしたいパイロットは，「cancel IFR」の用語を使えばよい．

PIL:　Cleared for VOR Z runway 06 approach via Melos South Arrival, cross YACHI at or above 6,000, Nippon Air 155.

PIL:　Sapporo Control, Nippon Air 155, request traffic information around Aomori airport.

ACC:　Nippon Air 155, expect one departure traffic, Tikyu Three Departure to New Chitose airport.

PIL:　Nippon Air 155, roger, this time, cancel IFR, make simulated VOR Z runway 06 approach via Melos South Arrival to Aomori airport.

ACC:　Nippon Air 155, roger, squawk VFR, Aomori QNH 2990, stand by frequency change.

PIL:　Squawk VFR, 2990, Nippon Air 155.

ACC:　Nippon Air 155, contact Aomori Tower 118.3.

PIL:　Aomori Tower 118.3, Nippon Air 155.

PIL:　Aomori Tower, Nippon Air 155.

TWR:　Nippon Air 155, Aomori Tower, go ahead.

PIL:　Nippon Air 155, VFR, now making simulated VOR Z runway 06 approach via Melos South Arrival, position departed YACHI.

TWR:　Nippon Air 155, runway 06, wind 020 at 3, QNH 2990, report MELOS.

Phraseology Example 10

　目視進入とは，計器飛行方式により飛行する航空機が行う進入の方法であり，計器進入方式の全部又は一部を所定の方式によらないで飛行場を視認しながら行うものである．

　なお，管制機関等から目視進入を行うように要求されたり示唆されることはなく，パイロットの要求に基づいてのみ許可される．また，VFR 機との間隔設定と地上障害物の回避は操縦士の責任で行う．（P.65 参照）

PIL: Miho Radar, Nippon Air 3651, maintain 10,000.

RDR: Nippon Air 3651, Miho Radar, expect localizer Y runway 25 approach, Izumo runway 25, wind 200 at 13, QNH 2981, recleared direct Izumo, descend and maintain 7,000.

PIL: QNH 2981, direct Izumo, descend 7,000, Nippon Air 3651.

PIL: Miho Radar, Nippon Air 3651, after leaving high station, request contact approach.

RDR: Nippon Air 3651, roger, will advise later, descend and maintain 4,000.

PIL: Descend 4,000, Nippon Air 3651.

RDR: Nippon Air 3651, cleared for localizer Y runway 25 approach, maintain 4,000, cross high station at 4,000, and also cleared for contact approach upon reaching Izumo.

PIL: Nippon Air 3651, cleared for localizer Y runway 25 approach, maintain 4,000, cross high station at 4,000, and cleared for contact approach upon reaching Izumo.

RDR: Nippon Air 3651, contact Izumo Radio 122.7.

PIL: 122.7, Nippon Air 3651.

PIL: Izumo Radio, Nippon Air 3651, we have approach clearance, after leaving high station, make contact approach, right downwind for runway 25.

AFIS: Nippon Air 3651, Izumo Radio, runway 25, QNH 2981, report right downwind.

　飛行場対空援助業務が実施される空港等で計器進入方式の種類を指定しない進入（cleared for approach）が許可されている場合は，レディオに対しては目視進入を行う旨の通報で足りるが，タワーがある場合は，管制官が交通状況を考慮の上，当該管制所が許可（cleared for contact approach）を行う．

　計器進入方式が指定された進入許可（cleared for ILS Z runway *** approach 等）が発出されている場合は，進入管制業務を行う機関に承認（cleared for contact approach）を要求し，その後，レディオに対しては行う旨を通報，タワーに対しては許可を受ける必要がある．

（参考２） Basic Example で扱っている大阪空港において，32L に進入し，その後 32R へ着陸する方法は，形の上では周回進入として扱われるので，「circle to runway 32R」の用語が使用される．

PIL:	Kansai Approach, Nippon Air 82, leaving 156 for 13,000, with I, request landing runway 32R.
APP:	Nippon Air 82, Kansai Approach, roger, expect runway 32R, recleared direct IKOMA.
PIL:	Recleared direct IKOMA, Nippon Air 82.
APP:	Nippon Air 82, descend and maintain 6,000.
PIL:	Descend and maintain 6,000, Nippon Air 82.
APP:	Nippon Air 82, descend and maintain 4,000.
PIL:	Descend and maintain 4,000, Nippon Air 82.
APP:	Nippon Air 82, 5 miles from IKOMA, descend and maintain 3,500, cleared for ILS runway 32L approach, circle to runway 32R, contact Osaka Tower 118.1.
PIL:	Descend 3,500, cleared for ILS runway 32L approach, circle to runway 32R, Tower 118.1, Nippon Air 82.
PIL:	Osaka Tower, Nippon Air 82, runway 32R.
TWR:	Nippon Air 82, Osaka Tower, report right break, wind 210 at 6.
PIL:	Report right break, Nippon Air 82.
PIL:	Osaka Tower, Nippon Air 82, right break.
TWR:	Nippon Air 82, runway 32R cleared to land, wind 210 at 7.

UNIT.6. Landing & Taxiing Back

Words & Phrases

after completing low approach (touch and go, etc...), maintain VMC

　ローアプローチ（タッチアンドゴー等）終了後，VMC を維持して下さい

after completing low approach (touch and go, etc...), turn left (right) heading / fly heading / continue runway heading ~, climb and maintain ~

　ローアプローチ（タッチアンドゴー等）終了後，左／右旋回／磁針路 ~ ／滑走路の方位で飛行，上昇して~を維持して下さい

after completing touch and go (stop and go, option), execute ~, climb via SID to ~

　タッチアンドゴー（ストップアンドゴー，オプションアプローチ）終了後，~により飛行し，SID の制限に従い，~まで上昇して下さい

report out of ILS critical area	ILS critical area not protected
ILS 制限区域の離脱を通報して下さい	ILS 制限区域は保護されていません
go around (*1)	execute missed approach (*2)
復行して下さい	進入復行して下さい

＊（*1）復行（go around）とは，着陸又はそのための進入の継続を中止して上昇体勢に移ることをいう.
＊（*2）進入復行（missed approach）とは，計器進入中の航空機が計器進入の継続を中止し，公示又は事前に通報された進入復行方式に従って飛行することをいう. なお，進入復行方式（missed approach procedure）とは計器進入が継続できない場合に航空機が従う飛行方式をいう.

Introduction

　進入開始後は MAPt まで進入を継続し，DA（非精密進入にあっては MDA での VDP），着陸の最低気象条件を満たしているか否かを判断して，「着陸に必要な条件」を満たさないと判断した場合は復行を行う.

Basic Example

飛行場管制所（Osaka Tower）・飛行場管制所地上管制席（Osaka Ground）と交信

> **PIL:** Osaka Tower, Nippon Air 710, runway 32L.
>
> **TWR:** Nippon Air 710, Osaka Tower, number 2 to runway 32L, report 6 DME, wind 210 at 6.
>
> **PIL:** Report 6 DME, Nippon Air 710.
>
>
> **PIL:** Osaka Tower, Nippon Air 710, 6 DME.
>
> **TWR:** Nippon Air 710, traffic vacating, runway 32L cleared to land, wind 200 at 6.
>
> **PIL:** Runway 32L cleared to land, Nippon Air 710.
>
>
> **TWR:** Nippon Air 710, turn right W-8, cross runway 32R.
>
> **PIL:** Nippon Air 710, turn right W-8, cross runway 32R.
>
> **TWR:** Nippon Air 710, cross runway without delay, traffic to 32R 6 miles on final, contact Ground 121.7.
>
> **PIL:** 121.7, without delay, Nippon Air 710.
>
>
> **PIL:** Osaka Ground, Nippon Air 710, crossing runway 32R, spot 9.
>
> **GND:** Nippon Air 710, Osaka Ground, taxi via A, hold short of E-3.
>
> **PIL:** Taxi via A, hold short of E-3, Nippon Air 710.
>
>
> **GND:** Nippon Air 710, after clearing 777 left side, taxi via E-3 to spot 9.
>
> **PIL:** After clearing 777, taxi via E-3 to spot 9, Nippon Air 710.

PART 1.

INSTRUMENT APPROACH CHART

RJOO / OSAKA INTL   ILS RWY32L

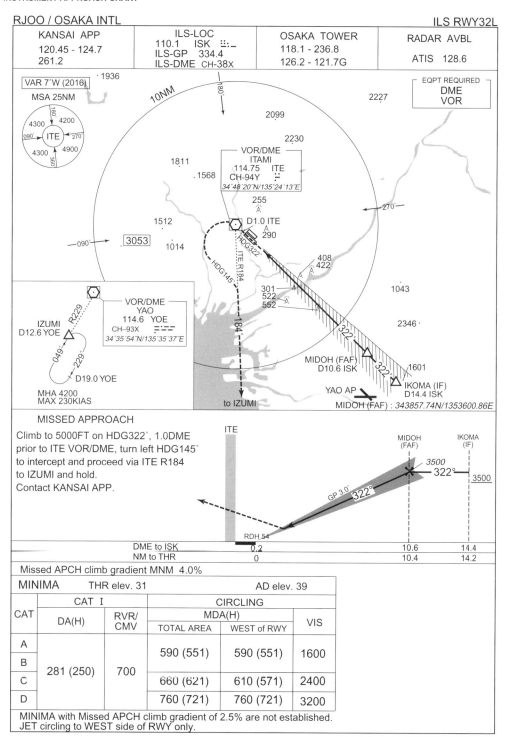

KANSAI APP 120.45 - 124.7 261.2	ILS-LOC 110.1 ISK ⠶⠒ ILS-GP 334.4 ILS-DME CH-38X	OSAKA TOWER 118.1 - 236.8 126.2 - 121.7G	RADAR AVBL ATIS 128.6

MISSED APPROACH

Climb to 5000FT on HDG322°, 1.0DME prior to ITE VOR/DME, turn left HDG145° to intercept and proceed via ITE R184 to IZUMI and hold.
Contact KANSAI APP.

		0.2	10.6	14.4
DME to ISK		0.2	10.6	14.4
NM to THR		0	10.4	14.2

Missed APCH climb gradient MNM 4.0%

MINIMA	THR elev. 31		AD elev. 39		
CAT	CAT I		CIRCLING		
	DA(H)	RVR/ CMV	MDA(H)		VIS
			TOTAL AREA	WEST of RWY	
A	281 (250)	700	590 (551)	590 (551)	1600
B			590 (551)	590 (551)	1600
C			660 (621)	610 (571)	2400
D			760 (721)	760 (721)	3200

MINIMA with Missed APCH climb gradient of 2.5% are not established.
JET circling to WEST side of RWY only.

Phraseology Example 1

着陸許可は，滑走路番号を前置して発出され，それに引き続き風向風速が通報される．パイロットは，滑走路番号もリードバックするべきである．なお，タワーへの通信設定については現在位置を慣例的に通報する場合がある．

滑走路を離脱した後は，管制機関の指示に従って地上走行を行う．

PIL:	Miyazaki Tower, Nippon Air 619, over SINWA.
TWR:	Nippon Air 619, Miyazaki Tower, runway 27 cleared to land, wind 350 at 6.
PIL:	Runway 27 cleared to land, Nippon Air 619.
TWR:	Nippon Air 619, turn left S-3, contact Ground 121.9.
PIL:	Turn left S-3, contact Ground 121.9, Nippon Air 619.
PIL:	Miyazaki Ground, Nippon Air 619, S-3, spot 5.
GND:	Nippon Air 619, Miyazaki Ground, taxi to spot 5.
PIL:	Taxi to spot 5, Nippon Air 619.

飛行場対空援助業務が実施される空港の場合は，「runway is clear」の用語が使用される．

PIL:	Fukushima Radio, Nippon Air 81, departed SOUMA.
AFIS:	Nippon Air 81, runway 01 runway is clear, wind 220 at 2 knots.
PIL:	Runway is clear, runway 01, Nippon Air 81.
AFIS:	Nippon Air 81, taxi to spot 1.
PIL:	Taxi to spot 1, Nippon Air 81.

離陸時の離陸時刻通報の場合と同様に，着陸時刻や滑走路離脱の報告を求められる場合がある．

PIL: Tottori Radio, Nippon Air 291, base turn inbound.

AFIS: Nippon Air 291, roger, runway 10 runway is clear, wind 160 degrees at 4 knots, report downtime.

PIL: Runway 10 runway is clear, report downtime, Nippon Air 291.

PIL: Tottori Radio, Nippon Air 291, request surface wind.

AFIS: Wind 140 degrees at 4 knots.

PIL: Thank you.

PIL: Tottori Radio, Nippon Air 291, down at 43.

AFIS: Nippon Air 291, down at 43 copied, report runway vacated.

PIL: Report runway vacated, Nippon Air 291.

PIL: Tottori Radio, Nippon Air 291, runway vacated.

AFIS: Nippon Air 291, roger, see you.

Phraseology Example 2

　パイロットが復行を行う場合には，そのときの位置に関係なく速やかに管制機関に通報しなければならない.

PIL:	Tokyo Tower, Nippon Air 652, approaching APOLO, spot 7.
TWR:	Nippon Air 652, Tokyo Tower, runway 34L cleared to land, wind 320 at 17, Boeing 787 5 miles on final.
PIL:	Cleared to land, runway 34L, Nippon Air 652.
TWR:	Nippon Air 652, microburst alert, runway 34L arrival, 44 knots loss, 3 miles final, use caution.
PIL:	Roger.
PIL:	Tokyo Tower, Nippon Air 652, go around.
TWR:	Nippon Air 652, turn left heading 220, climb and maintain 4,000.
PIL:	Turn left heading 220, climb and maintain 4,000, Nippon Air 652.
TWR:	Nippon Air 652, contact Departure 126.0.

　なお，「go around」の通報と一緒に，自らの要求を通報して許可を得る場合がある.

PIL:	Nippon Air 710, go around, request downwind.
TWR:	Nippon Air 710, roger, take downwind.
又は,	
PIL:	Nippon Air 710, go around, missed approach.
TWR:	Nippon Air 710, roger, go around, execute missed approach.

　滑走路又は航空交通状況等により進入継続が安全でないと管制機関が判断した場合も復行が指示される. なお，以後の飛行方法については，適切な時機に指示がある.

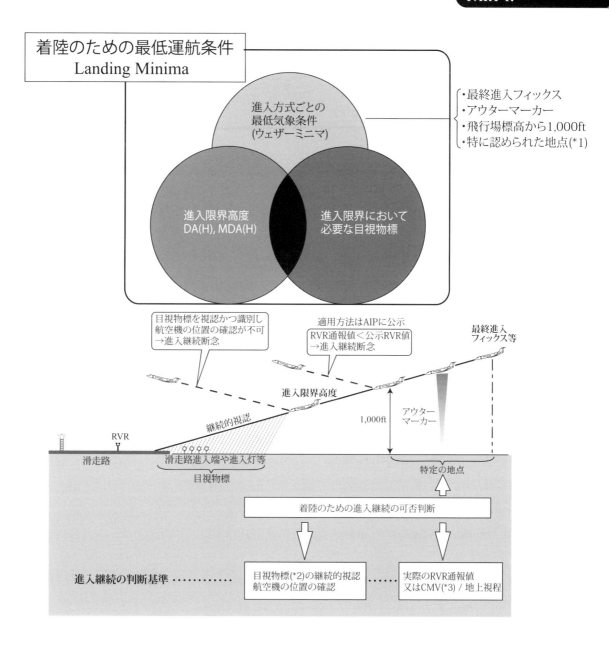

* （*1）航空会社が運航規程で定めて認可を受けた点.
* （*2）AIP に公示されている. 例えば, ILS CAT I / PAR / 非精密進入の場合は以下のうち
少なくとも１つ.

        １. 進入灯の一部   ２. 滑走路進入端   ３. 滑走路進入端標識

        ４. 滑走路末端灯   ５. 滑走路末端識別灯   ６. 進入角指示灯

        ７. 接地帯又は接地帯標識   ８. 接地帯灯   ９. 滑走路灯

        １０. 進入灯と同時運用されている直線進入用進入路指示灯   １１. 指示標識

* （*3）ILS CAT I（APV Baro-VNAV 進入を含む）と非精密直線進入においてのみ, RVR
が使用できないとき（RVR 2,000 m 超の場合を含む）に CMV（Converted Meteorological
Visibility：地上視程換算値）が適用される（P.81 参照）.

Phraseology Example 3

go around 後，再び進入を行う場合は，以下のようになる．

PIL:	Contact Tokyo Departure 126.0, Nippon Air 652.
PIL:	Tokyo Departure, Nippon Air 652, leaving 3,100 climbing 4,000, heading 220.
DEP:	Nippon Air 652, Tokyo Departure, radar contact, continue present heading, maintain 4,000.
PIL:	Nippon Air 652, continue present heading, maintain 4,000.
DEP:	Nippon Air 652, report your intention, another approach or divert.
PIL:	Nippon Air 652, we'd like another approach.
DEP:	Nippon Air 652, roger, expect ILS Z runway 34L approach.
PIL:	Nippon Air 652.
DEP:	Nippon Air 652, turn left heading 180.
PIL:	Left 180, Nippon Air 652.
DEP:	Nippon Air 652, contact Tokyo Approach 119.1.
PIL:	Contact Tokyo Approach 119.1, Nippon Air 652.
PIL:	Tokyo Approach, Nippon Air 652, 4,000, heading 180.
APP:	Nippon Air 652, Tokyo Approach, climb and maintain 5,000, expect vector to ARLON.

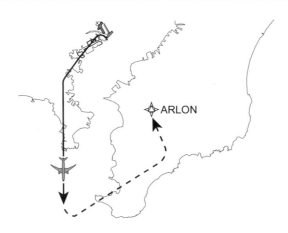

Phraseology Example 4

他の飛行場へダイバートする場合は，以下のようになる.

PIL: Tokyo Departure, Nippon Air 652.

DEP: Nippon Air 652, go ahead.

PIL: Nippon Air 652, destination change to Chubu Centrair, request radar vector.

DEP: Nippon Air 652, roger, copy, continue present heading, climb and maintain 4,000.

PIL: Continue present heading, 4,000, Nippon Air 652.

DEP: Nippon Air 652, turn right heading 250.

PIL: Right heading 250, Nippon Air 652.

DEP: Nippon Air 652, climb and maintain FL 200.

PIL: Climb FL 200, Nippon Air 652.

DEP: Nippon Air 652, clearance, cleared to Chubu Centrair airport via radar vector, fly heading 250, maintain FL 200.

✈ POINT: 地上視程から CMV への換算

　CMV は「観測された地上視程」に下表の「倍率」を掛けることで求められる. なお, RVR が適用される最大値は 2,000 m である.

　例えば, 夜間にあって進入灯及び滑走路灯が運用されている場合, 地上視程通報値が 1,200 m であれば, CMV=1,200 m × 2.0=2,400 m となる. このとき, 最低気象条件が RVR/ CMV=1,600 m であれば, 2,400 m ＞ 1,600 m として, Above Minima と判断される.

運用されている航空灯火	昼間	夜間
進入灯及び滑走路灯	倍率１．５	倍率２．０
滑走路灯	倍率１．０	倍率１．５
上記以外	倍率１．０	適用なし

計器進入には，大きく分けて精密進入方式と非精密進入方式がある．

精密進入（precision approach）とは，計器飛行による進入であって，進入方向（azimuth）と降下経路（glidepath）の情報・指示を受けることができるものをいう．非精密進入とは，精密進入以外の計器進入を指す．

ILS 進入の場合であって，グライドスロープ使用不能時のローカライザーアプローチ（localizer approach）は降下経路の指示を受けることができないので非精密進入となり，進入許可は「cleared for ILS runway ~ approach, glide slope out of service」のようになる（もともと「ローカライザー単独方式」が設定されている場合は，「cleared for localizer approach」の用語が使用される）．

計器進入

精密進入

ILS approach
PAR approach

Azimuth(進入方向)
と Glide Slope(降下経路)の指示

非精密進入

LOC approach
VOR(VOR/DME)approach
NDB(ADF)approach
TACAN approach
RNP approach
RNP AR approach
Surveillance approach

Azimuth(進入方向)
の指示

なお，以下のことに留意する．

・ローカライザー機能が停止した場合→グライドスロープ，ILS の DME も停止するので，計器進入方式としては成り立たない．

・ILS に DME が併設されている場合→グライドスロープ機能が停止されても，DME は同時に停止されない．

・標準式進入灯が不作動の場合は，Landing Minima を引き上げて，ILS 進入を行うことができる．

Azimuth（進入方向）
LOC Antenna

Glide Slope（降下経路）
GP Antenna
Glidepath

DH
滑走路進入端標高
DA
PALS
平均海面

✈ POINT:　DA (Decision Altitude) と DH (Decision Height)

DA（決心高度）/ DH（決心高）

精密進入又は垂直方向ガイダンス付き進入において，進入継続に必要な目視物標をその到達時に視認できない場合は進入復行を開始しなければならない高度／高さをいう．

なお，決心高度は平均海面を基準とし，決心高は滑走路末端標高又は接地帯標高を基準とする．

　非精密進入（non-precision approach）とは，計器進入の最終進入を水平方向のみの航法情報により行う計器進入である．グライドパス情報が得られないため，進入限界高度はMDA/Hによって示される．非精密進入では一般に，最終進入開始点からMDAまでの降下はパイロットの判断で行い，MDAを維持して着陸の可否を判断し，目視降下点（VDP）か，それに相当する地点から目視による着陸の降下を行う．滑走路若しくはvisual referenceを視認できない場合には，「go around」を通報し，進入復行点（MAPt）においてmissed approach procedureを実施する．なお，非精密進入の場合は通常，MAPtは局上となる．局が飛行場から離れているケースではDME値や局からの飛行時間により表示され，滑走路末端が原則である．

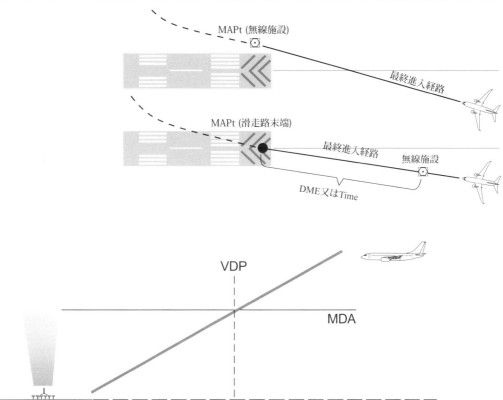

✈POINT:　MDA（Minimum Descent Altitude），MDH（Minimum Descent Height）とVDP（Visual Descent Point）

MDA（最低降下高度）/ MDH（最低降下高）

　非精密進入及び周回進入において定める，進入継続に必要な目視物標を視認することなくそれ未満へ降下してはならない高度／高さをいう．

　なお，最低降下高度は平均海面を基準とし，最低降下高は飛行場標高又は滑走路末端標高を基準とする．

VDP（目視降下点）

　非精密進入により直線進入を通常降下により行う場合において，進入灯又は滑走路末端を識別できる視覚援助施設を視認できたときに，最低降下高度以下に降下を開始する位置をいう．

　ILS 進入に必要な施設で，CAT I ILS アプローチが運用されるためには次の施設が必要である．

　　1．localizer（ローカライザー）

　　2．glide slope（グライドスロープ）

　　3．outer marker（アウターマーカー）又は DME

　　4．middle marker（ミドルマーカー）又は DME

　　5．PALS（標準式進入灯）

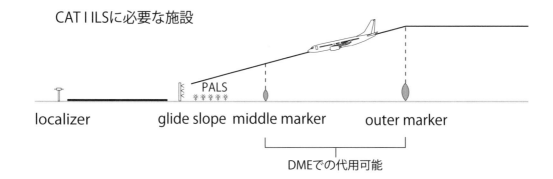

CAT I ILSに必要な施設

ILS 進入のカテゴリー

	DH	RVR	空港 (*1)
CAT I	200 ft 以上	550 m 以上 又は 地上視程 800 m 以上	ほとんどの空港 (*2)
CAT II	100 ft 以上 200 ft 未満	300 m 以上	関空 （06L/06R/24L/24R ILS） 中部 （18 ILS）
CAT III	100 ft 未満 又は定めなし	300 m 未満 50 m 以上	新千歳 （19R ILS）・釧路 （17 ILS） 青森 （24 ILS） 成田 （16R ILS）・羽田 （34R ILS） 中部 （36 ILS）・広島 （10 ILS） 熊本 （07 ILS）

＊（*1）2023 年 2 月現在の設置状況である．

＊（*2）多くの日本の空港では，1 本の滑走路の少なくとも片側に設置されている場合が多い．

PART.2.

IFR Communications - 2

PART.1. では，無線交信における一般的なやりとり・流れを扱ったが，PART. 2. では，操縦訓練を想定した ATC を取り上げる．大体の流れは PART.1. の通りであり，PART. 2. では説明等は最低限にしてある．会話の内容が不明な箇所は PART.1. に立ち返り再度確認して頂きたい．

Basic Example では，仙台空港から花巻空港へ向かう航空機を取り上げる．

Basic Example

Aircraft Identification	JA 5807
Type of Aircraft	Beechcraft G58 Baron
Departure Aerodrome	Sendai airport
Level	9,000 feet
Route	SDE - V-33 - HPE
Destination Aerodrome	Hanamaki airport
SID	SENDAI REVERSAL SIX DEP
Instrument Approach Procedure	VOR RWY 20 APP

UNIT.1. Departures

Basic Example

飛行場管制所地上管制席（Sendai Ground）と交信

PIL: Sendai Ground, JA 5807, information C.

GND: JA 5807, Sendai Ground, go ahead.

PIL: JA 5807, at CAC, request taxi to R and IFR clearance to Hanamaki airport, proposing 9,000.

GND: JA 5807, roger, runway 27, taxi to R area, stand by clearance.

PIL: Runway 27, taxi to R, stand by clearance, JA 5807.

GND: JA 5807, Sendai Ground, clearance.

PIL: JA 5807, go ahead.

GND: JA 5807 cleared to Hanamaki airport via Sendai Reversal Six Departure, flight planned route, maintain 9,000, squawk 4776.

PIL: JA 5807 cleared to Hanamaki airport via Sendai Reversal Six Departure, flight planned route, maintain 9,000, squawk 4776.

GND: JA 5807, read back is correct.

＊ CAC とは，航空大学校（Civil Aviation College）の（エプロンの）ことで，R とは run-up area のことである．

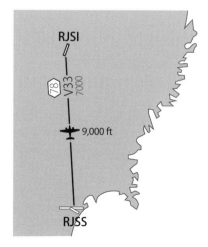

Phraseology Example 1

目的飛行場において Local Training 等を行う場合は，Y-Plan で飛行計画を作成する場合
が多い．クリアランスリミットがフィックスの場合は，以下のようになる．

GND:　JA 5809, clearance.

PIL:　　JA 5809, go ahead.

GND:　JA 5809 cleared to MORIO via Sendai Reversal Six Departure, flight
　　　　 planned route, maintain 13,000, squawk 3445.

PIL:　　JA 5809 cleared to MORIO via Sendai Reversal Six Departure, flight
　　　　 planned route, maintain 13,000, squawk 3445.

GND:　JA 5809, read back is correct.

...Sendai (SDE) - V-22 - MIYAKO (MQE) - MORIO - RJSI

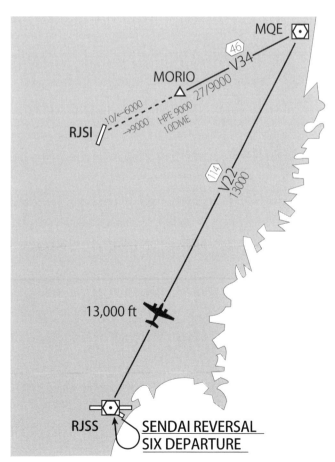

Phraseology Example 2

出発経路の指定（SID や Transition 等）は，離陸後の飛行方向，障害物，他の IFR 機との関係等の理由により決められるが，離陸後，レーダー誘導によって上昇する場合は，以下のようになる．

PIL:	Niigata Tower, JA 5810, at B-4, request IFR clearance to Sendai airport, proposing 11,000.
TWR:	JA 5810, Niigata Tower, stand by clearance.
PIL:	Stand by clearance, JA 5810.
TWR:	JA 5810, clearance.
PIL:	JA 5810, go ahead.
TWR:	JA 5810 cleared to Sendai airport via radar vector, R-217, Sendai, maintain 11,000, squawk 3411.
PIL:	JA 5810 cleared to Sendai airport via radar vector, R-217, Sendai, maintain 11,000, squawk 3411.
TWR:	JA 5810, read back is correct, Radar frequency 121.4, report when ready for departure.
PIL:	Radar frequency 121.4, report when ready, JA 5810.

...R-217 - Sendai (SDE)...

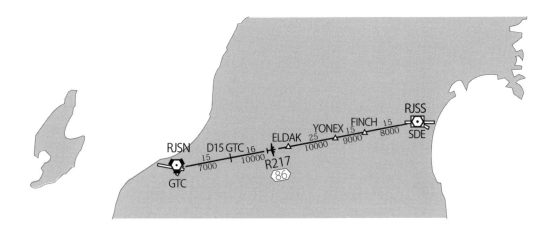

Phraseology Example 3

飛行場対空援助業務が実施される飛行場から出発する場合は，「ATC clears」の用語が使用される．なお，交通状況等により，地上待機の指示が発出される場合は，以下のようになる．

PIL:	Yamagata Radio, JA 5811, at spot A, request IFR clearance to Sendai airport, proposing 7,000.
AFIS:	JA 5811, Yamagata Radio, to Sendai, propose 7,000, temperature 28, QNH 2878, stand by for clearance.
PIL:	QNH 2878, stand by clearance, JA 5811.
AFIS:	JA 5811, clearance.
PIL:	JA 5811, go ahead.
AFIS:	ATC clears JA 5811 cleared to Sendai airport via Yamagata Four Departure, flight planned route, maintain 7,000, squawk 3321, hold on the ground.
PIL:	ATC clears JA 5811 cleared to Sendai airport via Yamagata Four Departure, flight planned route, maintain 7,000, squawk 3321, hold on the ground.
AFIS:	JA 5811, read back is correct.

Phraseology Example 4

交通状況等の理由（飛行場対空援助業務が実施されている空港等で，到着機を待たせずに先に出発機を出す場合等）により，管制承認失効時刻（clearance void time）が指定される場合もある．

AFIS:	JA 5806, vifno clearance.
PIL:	JA 5806, go ahead.
AFIS:	ATC clears JA 5806 cleared to Sendai airport via Fukushima Reversal Two Departure, flight planned route, maintain 7,000, squawk 4764, vifno 0640.
PIL:	ATC clears JA 5806 cleared to Sendai airport via Fukushima Reversal Two Departure, flight planned route, maintain 7,000, squawk 4764, vifno 0640.
AFIS:	JA 5806, read back is correct.

UNIT.2. Taxiing & Take-off

Basic Example

飛行場管制所（Sendai Tower）と交信

> **PIL:** Sendai Ground, JA 5807, at R, request B-5 intersection departure.
>
> **GND:** JA 5807, B-5 intersection approved, runway 27, taxi to holding point B-5 via D-1, cross runway 12.
>
> **PIL:** B-5 intersection approved, taxi to holding point B-5 via D-1, cross runway 12, JA 5807.
>
>
> **GND:** JA 5807, contact Tower 118.7.
>
> **PIL:** Contact Tower 118.7, JA 5807.
>
>
> **PIL:** Sendai Tower, JA 5807, taxiing to B-5, ready.
>
> **TWR:** JA 5807, Sendai Tower, wind 180 at 5, runway 27 at B-5, cleared for take-off.
>
> **PIL:** Runway 27 at B-5, cleared for take-off, JA 5807.
>
>
> **TWR:** JA 5807, contact Approach 120.4.
>
> **PIL:** Contact Approach 120.4, JA 5807.

AERODROME CHART

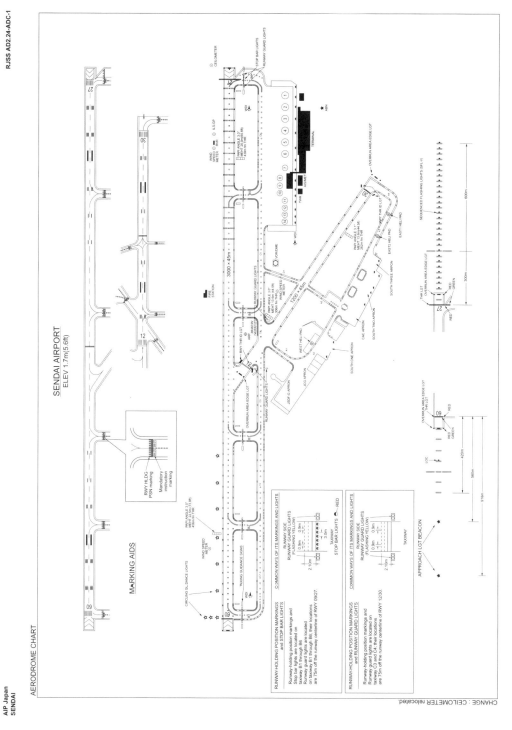

SENDAI AIRPORT
ELEV 1.7m(5.6ft)

CHANGE : CEILOMETER relocated

Phraseology Example 1

離陸を開始する前に離陸後のヘディングを指示されることがある.

PIL:	Niigata Tower, JA 5810, B-4, ready.
TWR:	JA 5810, continue runway heading, wind 150 at 8, runway 10 cleared for take-off.
PIL:	Runway 10 cleared for take-off, continue runway heading, JA 5810.
TWR:	JA 5810, contact Radar.
PIL:	Contact Radar, JA 5810.

Phraseology Example 2

出発制限の指示が取り消される場合は,「released for departure」の用語による.

PIL:	Yamagata Radio, JA 5811, ready, request taxi information and hold short of runway 01.
AFIS:	JA 5811, hold short of runway 01, wind 240 at 2, using runway 01, temperature 28, QNH 2978.
PIL:	Hold short of runway 01, QNH 2978, JA 5811.
AFIS:	JA 5811, Yamagata Radio, clearance.
PIL:	JA 5811, go ahead.
AFIS:	JA 5811, ATC clears JA 5811 released for departure.
PIL:	JA 5811, roger, ready.
AFIS:	JA 5811, backtrack runway 01, wind 130 at 2, report when ready.
PIL:	Backtrack runway 01, report when ready, JA 5811.
PIL:	JA 5811, ready.
AFIS:	JA 5811, traffic copter 5 miles north, use caution, wind 150 at 2, runway 01 runway is clear.
PIL:	Runway 01 runway is clear, JA 5811.

Phraseology Example 3

　管制承認失効時刻までに離陸しない場合は，当該管制承認は無効となる．管制承認失効時刻が付された場合であって，その時刻までに出発できないと判断した場合は，パイロットはその旨を通報するのが望ましい．

　仮に，管制承認失効時刻までに出発できない旨を通報した後は，その後代替指示として，以下のような指示や情報を受ける場合がある．

AFIS:　　JA 5806, revised clearance.

PIL:　　　JA 5806, go ahead.

AFIS:　　ATC clears JA 5806 cancel vifno, and hold on the ground.

PIL:　　　JA 5806, cancel vifno, and hold on the ground.

AFIS:　　JA 5806, read back is correct, traffic information, IFR inbound CRJ estimate Fukushima VOR 0653, VOR runway 19 approach.

PIL:　　　JA 5806, roger.

AFIS:　　JA 5806, revised clearance.

PIL:　　　JA 5806, go ahead.

AFIS:　　ATC clears JA 5806 released for departure.

UNIT.3. Climb

Basic Example

ターミナル管制所出域管制席（Sendai Approach）と交信

> PIL: Sendai Approach, JA 5807, leaving 1,300 assigned 9,000.
>
> APP: JA 5807, Sendai Approach, radar contact, fly heading 120 vector to Sendai VOR, climb and maintain 9,000.
>
> PIL: Fly heading 120, climb and maintain 9,000, JA 5807.
>
> APP: JA 5807, make left turn, resume own navigation direct Sendai VOR.
>
> PIL: Make left turn, resume own navigation direct Sendai VOR, JA 5807.
>
> APP: JA 5807, contact Tokyo Control 118.9.
>
> PIL: Contact Tokyo Control 118.9, JA 5807.

STANDARD DEPARTURE CHART-INSTRUMENT

RJSS / SENDAI   SID

SENDAI REVERSAL SIX DEPARTURE

RWY 09 : Climb RWY HDG to SDE 3.4DME (2.8NM fm DER), turn right to intercept
            and proceed...
RWY 12 : Climb ...
RWY 27 : Climb RWY HDG to 500FT, turn left HDG 090˚ to intercept and proceed...
RWY 30 : Climb RWY HDG to 500FT, turn left HDG 090˚ to intercept and proceed...
            ...via SDE R120 to 10.0DME, turn right, direct to SDE VOR/DME.
            Cross SDE VOR/DME at or above 7000FT(*).

*   In case of proceeding to IXE VOR/DME : Cross SDE VOR/DME at or
     above 5000FT.
     In case of proceeding to FKE VOR/DME : Cross SDE VOR/DME at or
     above 6000FT.

Note RWY 09 : 5.0% climb gradient required up to 500FT.
                    OBST ALT 62FT located at 0.2NM 102˚ FM end of RWY09.
        RWY 27 : 5.0% climb gradient required up to 1000FT.
                    OBST ALT 919FT located at 4.1NM 269˚ FM end of RWY27.
        RWY 30 : 5.0% climb gradient required up to 1200FT.
                    OBST ALT 1181FT located at 5.3NM 283˚ FM end of RWY30.

SENDAI REVERSAL SIX DEPARTURE

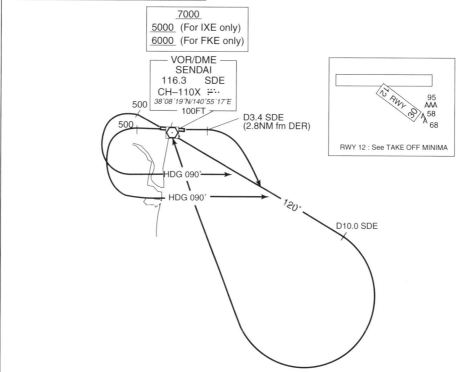

Phraseology Example 1

　管制機関（等）が，航空機に対して，トランスポンダーの識別（ident）機能の作動を指示して，航空機の位置（等）を確認することがある．

PIL:　　Sendai Approach, JA 5809, leaving 1,500 assigned 13,000.

APP:　　JA 5809, Sendai Approach, radar contact, maintain 3,000.

PIL:　　JA 5809, maintain 3,000.

APP:　　JA 5809, climb and maintain 5,000.

PIL:　　JA 5809, climb and maintain 5,000.

APP:　　JA 5809, climb and maintain 6,000.

PIL:　　JA 5809, climb and maintain 6,000.

APP:　　JA 5809, climb and maintain 13,000.

PIL:　　JA 5809, climb and maintain 13,000.

APP:　　JA 5809, ident for position confirmation, report present altitude.

PIL:　　JA 5809, leaving 10,000.

APP:　　JA 5809, ident observed, 4 miles north of Sendai VOR.

PIL:　　JA 5809.

APP:　　JA 5809, contact Tokyo Control 118.9.

PIL:　　JA 5809, contact Tokyo Control 118.9.

Phraseology Example 2

離陸後，レーダー誘導が開始される場合は，以下のようになる．

PIL: Niigata Radar, JA 5810, leaving 1,300 climbing 11,000.

RDR: JA 5810, Niigata Radar, radar contact, continue runway heading vector to R-217, maintain 11,000.

PIL: Continue runway heading, maintain 11,000, JA 5810.

RDR: JA 5810, report heading.

PIL: JA 5810, heading 100.

RDR: JA 5810, request VOR leaving.

PIL: JA 5810, now leaving.

RDR: JA 5810, resume own navigation, turn left heading 080 to R-217.

PIL: Left heading 080 to R-217, resume own navigation, JA 5810.

RDR: JA 5810, contact Tokyo Control 132.3.

PIL: Contact Tokyo Control 132.3, JA 5810.

Phraseology Example 3

　ACC により進入管制業務が行われている飛行場から出発する場合は，低高度ではレーダー管制業務が提供できない場合が多いため，SID 又は Transition を上昇中に，ACC と直接通信が可能な高度に達した時点で，ACC への周波数の切り替えが指示される場合がある．

PIL:　　Runway 01 runway is clear, JA 5811.

PIL:　　JA 5811, now passing 1,600.

AFIS:　JA 5811, after passing 4,000, contact Tokyo Control 118.9.

PIL:　　After passing 4,000, contact Tokyo Control 118.9, JA 5811.

PIL:　　Tokyo Control, JA 5811, leaving 4,500 climbing 7,000.

ACC:　　JA 5811, Tokyo Control, area QNH 2976, stand by radar pick up.

PIL:　　Area QNH 2976, stand by radar pick up, JA 5811.

ACC:　　JA 5811, ident.

PIL:　　Ident, JA 5811.

ACC:　　JA 5811, radar contact, position 15 miles north of Yamagata.

PIL:　　JA 5811, request direct Yamagata VOR.

ACC:　　JA 5811, recleared direct Yamagata VOR.

PIL:　　Recleared direct Yamagata VOR, JA 5811.

ACC:　　JA 5811, contact Sendai Approach 120.4.

---

after Yamagata VOR, proceed to Sendai VOR, mag course 128, zone distance 31, zone time 10, COP 10 from Yamagata VOR, MEA 7,000

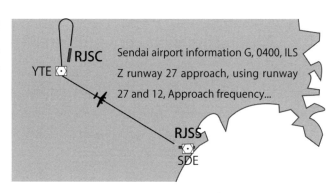

Sendai airport information G, 0400, ILS Z runway 27 approach, using runway 27 and 12, Approach frequency...

Phraseology Example 4

Phraseology Example 3 と同様，ACC とのイニシャルコンタクトにおいて，指定高度が 14,000 ft 未満の場合は，そのセクターごとにあらかじめ定められた地点の空域 QNH（area QNH）を入手する.

TWR: JA 5812, wind 120 at 3, runway 28 cleared for take-off.

PIL: JA 5812, runway 28 cleared for take-off.

TWR: JA 5812, contact Sapporo Control 127.57.

PIL: JA 5812, contact Sapporo Control 127.57.

PIL: Sapporo Control, JA 5812, leaving 1,600 assigned altitude 11,000.

ACC: JA 5812, Sapporo Control, area QNH 3013, squawk 2377.

PIL: JA 5812, area QNH 3013, squawk 2377.

ACC: JA 5812, radar contact, 1.5 miles west of YUWA VOR, report altitude.

PIL: JA 5812, leaving 3,000.

ACC: JA 5812, roger.

ACC: JA 5812, contact Tokyo Control 118.9.

PIL: JA 5812, contact Tokyo Control 118.9.

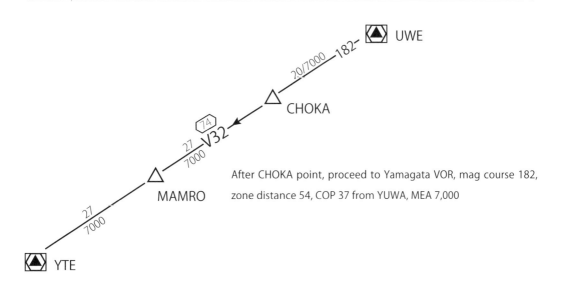

After CHOKA point, proceed to Yamagata VOR, mag course 182, zone distance 54, COP 37 from YUWA, MEA 7,000

UNIT.4. En-Route

Basic Example

管制区管制所（Tokyo Control）・管制区管制所（Sapporo Control）と交信

PIL:　　Tokyo Control, JA 5807, maintain 9,000.

ACC:　　JA 5807, Tokyo Control, area QNH 2998.

PIL:　　Area QNH 2998, JA 5807.

ACC:　　JA 5807, contact Sapporo Control 124.5.

PIL:　　Contact Sapporo Control 124.5, JA 5807.

PIL:　　Sapporo Control, JA 5807, maintain 9,000.

ACC:　　JA 5807, Sapporo Control, area QNH 2996.

PIL:　　Area QNH 2996, JA 5807.

ACC:　　JA 5807, descend at pilot's discretion maintain 8,000.

PIL:　　Descend at pilot's discretion maintain 8,000, JA 5807.

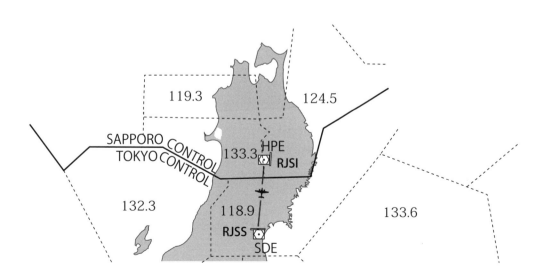

Phraseology Example 1

指定高度が 14,000 ft 未満の場合は，そのセクターごとにあらかじめ定められた地点又は経路上の空域 QNH（area QNH）を ACC より入手する．

> PIL: Tokyo Control, JA 5809, leaving 10,400 assigned 13,000.
>
> ACC: JA 5809, Tokyo Control, area QNH 2962.
>
> PIL: JA 5809, area QNH 2962.
>
> ACC: JA 5809, contact Sapporo Control 124.5.
>
> PIL: JA 5809, contact Sapporo Control 124.5.
>
> PIL: Sapporo Control, JA 5809, maintain 13,000.

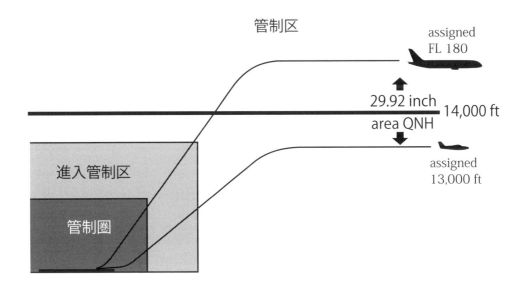

POINT: QNH－2

14,000 ft 未満であっても，以下の場合は QNH 値は提供されない．

1．管制区管制所が，FL 140 以上の高度を指定されている航空機と通信設定を行った場合

2．ターミナル管制所が，その管轄区域内にある飛行場から離陸し航空機と通信設定を行った場合

3．同一ターミナル管制機関内にて，継承機関が移管機関の QNH と同一観測点の地点のものを提供する場合

Phraseology Example 2

　新たな管制機関への通信移管が行われた場合，上昇中であれば通過高度（100 ft 単位）と指定高度を通報し，巡航中であれば指定高度を通報する．

PIL:　　Tokyo Control, JA 5810, leaving 6,300 climbing 11,000.

ACC:　　JA 5810, Tokyo Control, area QNH 2990.

PIL:　　Area QNH 2990, JA 5810.

ACC:　　JA 5810, contact Tokyo Control 118.9.

PIL:　　Contact Tokyo Control 118.9, JA 5810.

PIL:　　Tokyo Control, JA 5810, leaving 10,200 climbing 11,000.

ACC:　　JA 5810, Tokyo Control, area QNH 3003.

PIL:　　Area QNH 3003, JA 5810.

ACC:　　JA 5810, contact Sendai Approach 120.4.

---

airborne Niigata 38, maintain 11,000, estimate Sendai VOR 05 next

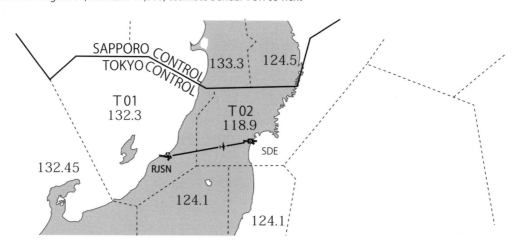

Sendai airport information J, 0700, RNP Z runway 09 approach, using
runway 09, Approach frequency 120.4, wind 120 degrees 10 knots, visibility
7 km, broken 3,000 cumulus, temperature 18, dew point 9, QNH 3005
inches, advise you have information J.

Phraseology Example 3

目的地へ接近した航空機は，付近のトラフィックの状況を判断し，必要があれば早めに降下の指示を得て，低い高度へ降下することもできる．

PIL:	Tokyo Control, JA 5812, maintain 11,000.
ACC:	JA 5812, Tokyo Control, area QNH 3010.
PIL:	JA 5812, area QNH 3010.
PIL:	Tokyo Control, JA 5812, request low altitude.
ACC:	JA 5812, descend and maintain 8,000.
PIL:	JA 5812, descend and maintain 8,000.
ACC:	JA 5812, contact Sendai Approach 120.4.
PIL:	JA 5812, contact Sendai Approach 120.4.

airborne Akita 0053, maintain 11,000, estimate Sendai VOR 0136.

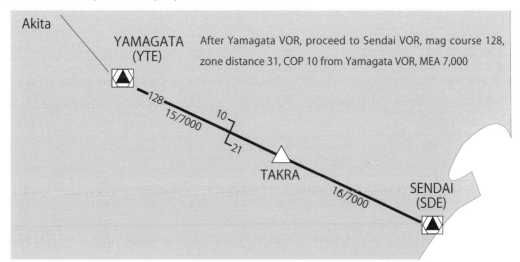

After Yamagata VOR, proceed to Sendai VOR, mag course 128, zone distance 31, COP 10 from Yamagata VOR, MEA 7,000

Sendai airport information E, 0100, RNP Z runway 09 approach, using runway 09, runway 12 and 30 closed, taxiway C-3, C-4 A-1 closed, Approach frequency 120.4, wind 130 degrees 11 knots, visibility 6 km, scattered 400 feet cumulus, temperature 19, dew point 11, QNH 3004 inches, advise you have information E.

UNIT.5. Descent & Approach

Basic Example

管制区管制所（Sapporo Control）・飛行場対空援助局（Hanamaki Radio）と交信

> **ACC:** JA 5807, maintain 7,000 or above until Hanamaki, cleared for VOR runway 20 approach.
>
> **PIL:** Maintain 7,000 or above until Hanamaki, cleared for VOR runway 20 approach, JA 5807.
>
>
> **ACC:** JA 5807, Hanamaki QNH 2993, contact Hanamaki Radio 118.2.
>
> **PIL:** QNH 2993, contact Hanamaki Radio 118.2, JA 5807.
>
>
> **PIL:** Hanamaki Radio, JA 5807, with 0000Z Hanamaki weather.
>
> **AFIS:** JA 5807, Hanamaki Radio, go ahead.
>
> **PIL:** JA 5807, we have approach clearance, estimate Hanamaki VOR 51, we will make VOR runway 20 approach.
>
> **AFIS:** JA 5807, runway 20, wind 290 at 6, QNH 2993, report high station.
>
> **PIL:** Runway 20, QNH 2993, report high station, JA 5807.

　ACC から計器進入を許可され，飛行場管制所又は飛行場対空援助局と通信を設定するよう指示された場合，レーダー業務は自動的に終了し，レーダー業務終了の通報は行われない（P.124 参照）。

　その場合，通常，パイロットは必要に応じて，

　　　１．計器進入方式の名称／進入許可を受領した旨

　　　２．その他（気象情報・高度・到着予定時刻等）

を通報するのが望ましい。

PART 2.

INSTRUMENT APPROACH CHART

RJSI / HANAMAKI   VOR RWY20

SAPPORO CONTROL 124.5 – 303.8 120.575 – 277.1	HANAMAKI VOR/DME 112.8 HPE CH–75X :–·–· 39°26′00″N/141°08′01″E	HANAMAKI RADIO 118.2 – 126.2	NO RADAR

VAR 8°W (2012)

MSA 25NM

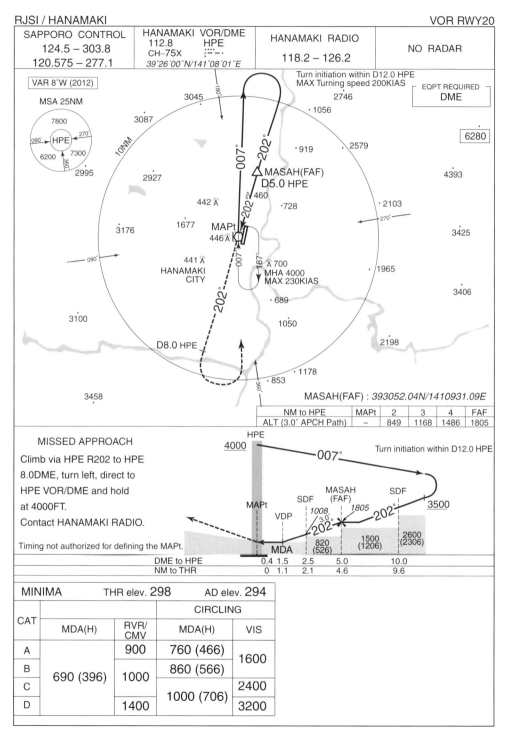

Turn initiation within D12.0 HPE
MAX Turning speed 200KIAS

EQPT REQUIRED
DME

6280

MASAH(FAF)
D5.0 HPE

MAPt

HANAMAKI CITY

MHA 4000
MAX 230KIAS

D8.0 HPE

MASAH(FAF) : 393052.04N/1410931.09E

NM to HPE	MAPt	2	3	4	FAF
ALT (3.0° APCH Path)	–	849	1168	1486	1805

MISSED APPROACH

Climb via HPE R202 to HPE
8.0DME, turn left, direct to
HPE VOR/DME and hold
at 4000FT.
Contact HANAMAKI RADIO.

Timing not authorized for defining the MAPt.

Turn initiation within D12.0 HPE

DME to HPE	0.4	1.5	2.5	5.0	10.0
NM to THR	0	1.1	2.1	4.6	9.6

MINIMA	THR elev. 298		AD elev. 294	
CAT			CIRCLING	
	MDA(H)	RVR/CMV	MDA(H)	VIS
A		900	760 (466)	1600
B	690 (396)	1000	860 (566)	
C			1000 (706)	2400
D		1400		3200

Phraseology Example 1

気象条件等の理由により VFR から IFR への変更を行う場合は，必要に応じて，

    1．IFR クリアランスの要求

    2．現在位置

    3．高度

等を適宜通報して，管制機関にクリアランスを要求する（最寄の管制機関との通信が困難である場合は，レディオ等を通じて要求する）．

PIL: Sapporo Control, JA 5808, VFR.

ACC: JA 5808, Sapporo Control, go ahead.

PIL: JA 5808, 14 miles east of YUWA VOR, leaving 4,300 climbing 7,500, request IFR clearance to Hanamaki airport via present position direct Hanamaki direct, proposing 7,000.

ACC: JA 5808, squawk 2105.

PIL: Squawk 2105, JA 5808.

ACC: JA 5808, radar contact, 18 miles east of YUWA, report altitude.

PIL: JA 5808, leaving 7,200 climb to 7,500.

ACC: JA 5808, roger, copy.

ACC: JA 5808, clearance.

PIL: JA 5808, go ahead.

ACC: JA 5808 cleared to Hanamaki airport via present position direct Hanamaki direct, maintain 7,000.

PIL: JA 5808 cleared to Hanamaki airport via present position direct Hanamaki VOR direct, maintain 7,000.

ACC: JA 5808, read back is correct, area QNH 3037, and request type of approach.

PIL: 3037, and request ILS Z runway 20 approach, JA 5808.

ACC: JA 5808, roger, intention copy.

ACC: JA 5808, cleared for ILS Z runway 20 approach, cross Hanamaki at or above 7,000.

PIL: Cleared for ILS Z runway 20 approach, cross Hanamaki at or above 7,000, JA 5808.

ACC: JA 5808, and contact Hanamaki Radio.

PIL: Contact Hanamaki Radio, JA 5808.

PIL: Hanamaki Radio, JA 5808.

AFIS: JA 5808, Hanamaki Radio, go ahead.

PIL: JA 5808, maintain 7,000, make ILS Z runway 20 approach from high station, we have 0200Z Hanamaki weather.

Phraseology Example 2

クリアランスリミットがフィックスの場合であって，フィックス到達前に IFR をキャンセルする場合は，以下のようになる.

PIL: Contact Tokyo Control 132.3, JA 5813.

PIL: Tokyo Control, JA 5813, 11,000.

ACC: JA 5813, Tokyo Control, area QNH 3015.

PIL: QNH 3015, JA 5813.

PIL: Tokyo Control, JA 5813, cancel IFR at 0119.

ACC: JA 5813, squawk VFR, frequency change approved.

PIL: Squawk VFR, frequency change approved, JA 5813.

PIL: Shonai Radio, JA 5813, VFR, we have 0100Z weather.

AFIS: JA 5813, Shonai Radio, go ahead.

PIL: JA 5813, 9 miles southeast of Shonai VOR, 9,000, we'll make simulated ILS Y runway 09 approach.

AFIS: JA 5813, roger, using runway 09, wind 100 at 8, QNH 3015, temperature 8, traffic not reported, report high station.

Phraseology Example 2

...SDE TAKRA YTE YSE / N0150VFR VFR LCL

Phraseology Example 3

　クリアランスリミット以降，IFR をキャンセルする予定であったが，気象条件等によりクリアランスリミット以降の IFR クリアランスを得る場合，以下のようになる．

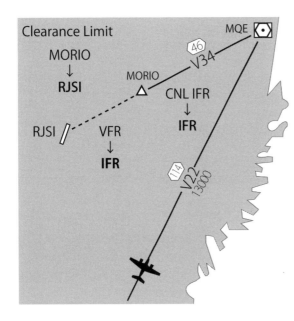

request IFR (approach) clearance to Hanamaki airport

Hanamaki 0300 weather: wind 050 degrees at 5 knots, direction variable between 020 and 090, visibility 30 km, cloud few 3,000 cumulus, temperature 17, dew point 11, QNH 2962, ...

clearance limit; MORIO

cancel IFR, make simulated ILS Y RWY 20 approach

airborne Sendai airport 0307, estimate MORIO 0400

飛行方式				（気象状況等により）
IFR	I-Plan	⇒	CNL IFR	
VFR	V-Plan	⇒		IFR Clearance を要求
IFR	Y-Plan	⇒	CNL IFR の前に	（Clearance Limit 以降の）IFR Clearance を要求
IFR	Y-Plan	⇒	CNL IFR の後に	IFR Clearance を要求

PIL: Sapporo Control, JA 5809, maintain 13,000, estimate MORIO 0400.

ACC: JA 5809, Sapporo Control, area QNH 2961.

PIL: JA 5809, area QNH 2961, and expect cancel IFR before MORIO.

ACC: JA 5809, roger.

PIL: Sapporo Control, JA 5809, due to cloud, request IFR clearance to Hanamaki airport via MIYAKO direct MORIO then direct, and request ILS Y runway 20 approach, expect cancel IFR on final.

ACC: JA 5809, roger, cleared to Hanamaki airport via MIYAKO direct MORIO then direct, maintain 13,000, stand by further clearance, you are number 2 due to inbound traffic.

PIL: Cleared to Hanamaki airport via MIYAKO direct MORIO then direct, maintain 13,000, JA 5809.

ACC: JA 5809, upon reaching MORIO, expect hold, expect approach at 0400.

PIL: JA 5809, upon reaching MORIO, expect hold, EAT 0400.

ACC: JA 5809, descend and maintain 9,000.

PIL: JA 5809, descend and maintain 9,000.

ACC: JA 5809, this time, (cancel hold clearance), cleared for ILS Y runway 20 approach, contact Hanamaki Radio 118.2.

PIL: JA 5809, (cancel hold clearance), cleared for ILS Y runway 20 approach, contact Hanamaki Radio 118.2.

PIL: Hanamaki Radio, JA 5809.

AFIS: JA 5809, Hanamaki Radio, go ahead.

PIL: JA 5809, we have Hanamaki 0300 weather, and we have approach clearance, we will make ILS Y runway 20 approach and touch and go, expect cancel IFR on final.

AFIS: JA 5809, roger, Hanamaki runway 20, wind 290 degrees at 9 knots, temperature 17, QNH 2962, after touch and go, turn left, use east side of pattern, traffic not reported, report depart MORIO.

PIL: JA 5809, QNH 2962, report depart MORIO.

Phraseology Example 4

視認進入を行う場合は，以下のようになる.

PIL: Sendai Approach, JA 5810, maintain 11,000, information J, request low altitude and visual approach via radar vector to runway 09 north downwind.

APP: JA 5810, Sendai Approach, intention copied, stand by low altitude due to traffic.

PIL: Stand by, JA 5810.

APP: JA 5810, fly heading 110 vector to runway 09 north downwind, descend and maintain 9,000.

PIL: Fly heading 110, descend and maintain 9,000, JA 5810.

APP: JA 5810, turn left heading 080, descend and maintain 8,000.

PIL: Turn left heading 080, descend and maintain 8,000, JA 5810.

APP: JA 5810, turn left heading 070, descend and maintain 6,000.

PIL: Turn left heading 070, descend and maintain 6,000, JA 5810.

APP: JA 5810, descend and maintain 2,500.

PIL: Descend and maintain 2,500, JA 5810.

APP: JA 5810, turn left heading 360.

PIL: Turn left heading 360, JA 5810.

APP: JA 5810, turn left heading 330.

PIL: Turn left heading 330, JA 5810.

APP: JA 5810, descend and maintain 1,500.

PIL: Descend and maintain 1,500, JA 5810.

APP: JA 5810, turn left heading 310.

PIL: Turn left heading 310, JA 5810.

APP: JA 5810, airport 9 o'clock, 8 miles, report airport in sight.

PIL: Roger, JA 5810.

APP: JA 5810, turn left heading 280.

PIL: Turn left heading 280, JA 5810.

APP: JA 5810, airport 9 o'clock, 7 miles, report airport in sight.

PIL: Airport in sight, JA 5810.

APP: JA 5810, roger, cleared visual approach runway 09, join north downwind, contact Tower 118.7.

PIL: Cleared visual approach runway 09, join north downwind, contact Tower 118.7, JA 5810.

PIL: Sendai Tower, JA 5810, proceed to north downwind.

TWR: JA 5810, Sendai Tower, runway 09, report left base.

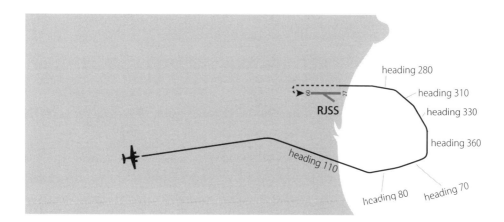

Phraseology Example 5

基礎旋回等を含む進入を行う場合は，以下のようになる．

PIL:　　Sendai Approach, JA 5811, maintain 7,000, information G.

APP:　　JA 5811, Sendai Approach, go ahead, request your intention.

PIL:　　JA 5811, request ILS Y runway 27 approach from high station.

APP:　　JA 5811, roger.

APP:　　JA 5811, maintain 7,000 until Sendai VOR, then cleared for ILS Y runway 27 approach, report high station.

PIL:　　Maintain 7,000 until Sendai VOR, then cleared for ILS Y runway 27 approach, report high station, JA 5811.

PIL:　　Sendai Approach, JA 5811, leaving high station.

APP:　　JA 5811, roger, report starting base turn.

PIL:　　Report starting base turn, JA 5811.

PIL:　　Sendai Approach, JA 5811, starting base turn.

APP:　　JA 5811, roger, contact Tower 118.7.

PIL:　　Contact Tower 118.7, JA 5811.

PIL:　　Sendai Tower, JA 5811, completing base turn.

TWR:　　JA 5811, Sendai Tower, runway 27 continue approach.

　精密進入の場合，進入許可受領後は，一般的には，「leaving high station」後は，通常「report starting base turn」又は「report establish localizer」等が指示され，その後，着陸許可が発出される．それ以外の場合においても，他のトラフィック等の状況により，「report 5 miles on final」のように Final course のある距離での通報を求められる場合もある．

　非精密進入の場合も精密進入の場合と同様であるが，「leaving high station」後は，通常「report starting base turn」又は「report completing base turn」を指示され，その後，着陸許可が発出される場合が多い．

なお，ホールディングが指示される場合は，以下のようになる.

PIL:	Tokyo Control, JA 5811, request direct TAKRA.
ACC:	JA 5811, recleared direct TAKRA.
PIL:	Recleared direct TAKRA, JA 5811.
ACC:	JA 5811, contact Sendai Approach 120.4.
PIL:	120.4, JA 5811.
PIL:	Sendai Approach, JA 5811, maintain 7,000, information G.
APP:	JA 5811, Sendai Approach, hold northwest of Sendai VOR, maintain 7,000.
PIL:	Hold northwest of Sendai VOR, maintain 7,000, JA 5811.
APP:	JA 5811, expect approach at 0250.
PIL:	EAT 0250, JA 5811.
PIL:	Sendai Approach, JA 5811, start holding at 39.
APP:	JA 5811, roger.
APP:	JA 5811, descend and maintain 4,000.
PIL:	Descend 4,000, JA 5811.
APP:	JA 5811, cleared for ILS Y runway 27 approach, report high station.

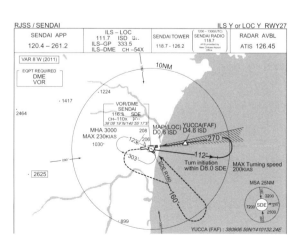

Phraseology Example 6

周回進入を行う場合は，以下のようになる．

PIL: Sendai Approach, JA 5810, maintain 11,000, information K, request ILS Z runway 27 approach via radar vector, circle to runway 09 approach.

APP: JA 5810, Sendai Approach, roger, stand by for coordination.

PIL: JA 5810.

APP: JA 5810, fly heading 110 vector to ILS Z runway 27 final approach course, descend and maintain 9,000.

PIL: Heading 110, descend 9,000, JA 5810.

APP: JA 5810, descend and maintain 4,000.

PIL: Descend 4,000, JA 5810.

APP: JA 5810, turn left heading 100.

PIL: Left heading 100, JA 5810.

APP: JA 5810, turn left heading 090.

PIL: Left heading 090, JA 5810.

APP: JA 5810, turn left heading 080, descend and maintain 1,500.

PIL: Left heading 080, descend 1,500, JA 5810.

APP: JA 5810, turn left heading 060.

PIL: Left heading 060, JA 5810.

APP: JA 5810, turn left heading 360.

PIL: Left heading 360, JA 5810.

APP: JA 5810, 4 miles from YUKKA, turn left heading 300, cleared for ILS Z runway 27 approach, circle to runway 09.

PIL: Left heading 300, cleared for ILS Z runway 27 approach, circle to runway 09, JA 5810.

APP: JA 5810, contact Tower 118.7.

PIL: Contact Tower 118.7, JA 5810.

PIL: Sendai Tower, JA 5810, 6 miles on final.

TWR: JA 5810, Sendai Tower, report right break for runway 09 circling approach.

PIL: Report right break, JA 5810.

PIL: Sendai Tower, JA 5810, right break.

TWR: JA 5810, report left base, runway 09.

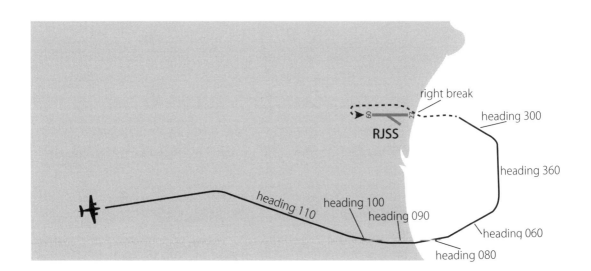

Phraseology Example 7

RNP 進入を行う場合は，以下のようになる．

PIL:	Sendai Approach, JA 5810, maintain 11,000, information J.
APP:	JA 5810, Sendai Approach, go ahead.
PIL:	JA 5810, request RNP Z runway 09 approach via radar vector.
APP:	JA 5810, roger, expect vector to SHIPS for RNP Z runway 09 approach.
PIL:	Expect vector to SHIPS for RNP Z runway 09 approach, JA 5810.
APP:	JA 5810, turn right heading 180 vector to SHIPS, you are number 3.
PIL:	Turn right 180 vector to SHIPS, number 3, JA 5810.
APP:	JA 5810, descend and maintain 9,000.
PIL:	Descend and maintain 9,000, JA 5810.
APP:	JA 5810, descend and maintain 8,000.
PIL:	Descend and maintain 8,000, JA 5810.
APP:	JA 5810, turn left heading 090, descend and maintain 7,000.
PIL:	Turn left heading 090, descend and maintain 7,000, JA 5810.
APP:	JA 5810, turn left heading 070, descend and maintain 6,000.
PIL:	Turn left heading 070, descend and maintain 6,000, JA 5810.
APP:	JA 5810, turn left heading 360.
PIL:	Turn left heading 360, JA 5810.
APP:	JA 5810, resume own navigation direct SHIPS.
PIL:	Resume own navigation direct SHIPS, JA 5810.

APP: JA 5810, descend and maintain 4,000, cross SHIPS at or above 4,000, cleared for RNP Z runway 09 approach.

PIL: Descend and maintain 4,000, cross SHIPS at or above 4,000, cleared for RNP Z runway 09 approach, JA 5810.

APP: JA 5810, contact Sendai Tower 118.7.

PIL: Contact Sendai Tower 118.7, JA 5810.

PIL: Sendai Tower, JA 5810, departed SHIPS.

TWR: JA 5810, Sendai Tower, runway 09, report SUGOH.

PIL: Runway 09, report SUGOH, JA 5810.

PIL: Sendai Tower, JA 5810, departed SUGOH.

TWR: JA 5810, runway 09 cleared to land, wind 110 at 6.

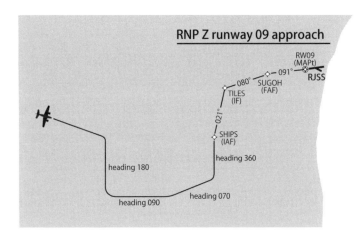

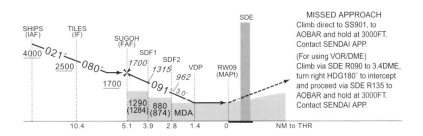

通常，管制官からの以下の指示（用語）により降下を行う．

    1．JA 5807, descend and maintain ***.

    2．JA 5807, descend at pilot's discretion maintain ***.

    3．JA 5807, descend to reach *** by ***.

    4．JA 5807, cleared for (~) approach.

しかしながら，降下の指示がなかった場合等は，降下を要求しなければならない．

    5．JA 5807, request descent (to 7,000) / request 7,000.

    6．JA 5807, request approach clearance.

    進入許可（上記４．の場合）が発出された後は，MEAまで降下を開始することができ（ホールディング中であればMHA），進入開始点以降への進入が可能である（missed approachまで含む）．進入許可は通常，進入開始点到達5分前までに発出される．

    なお，進入を開始するのに高度が高すぎる場合は，ホールディングパターン内等で高度処理を行わなければならない．

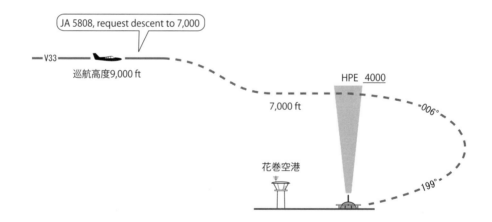

Phraseology Example 8

計器進入開始前に，公示又は自己の着陸最低気象条件を満たさない場合は，管制機関（等）にその旨を通報し，待機又は代替飛行場へのクリアランスを要求しなければならない．その際，通信機が故障した場合を考慮し，どの程度の時間ホールディングする予定であるかを併せて通報するのが望ましい．

> PIL: JA 5807, weather is below our landing minima. Request holding over Hanamaki VOR at 6,000 about 10 minutes for weather improvement.

天候により待機をしている場合（進入を行ったが MAPt で滑走路等を視認できなかった場合も含む）等で天候の回復が見込めない場合や機材トラブル等で，他空港・代替飛行場への飛行の変更を行う必要がある場合は，経路・高度を含む IFR クリアランスを要求する．

> PIL: Sapporo Control, JA 5807, request destination change to Sendai airport, request clearance to Sendai airport via present position direct Sendai VOR direct, proposing 8,000.
>
> ACC: JA 5807, roger, stand by.
>
>
> ACC: JA 5807, clearance, cleared to Sendai airport via present position direct Sendai VOR direct, maintain 8,000.

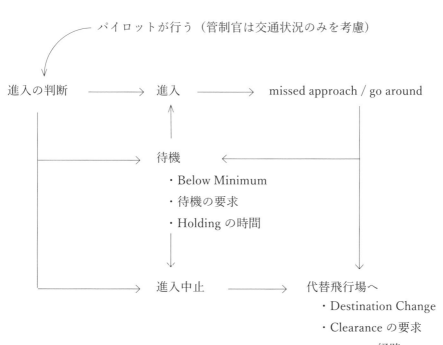

UNIT.6. Landing & Taxiing Back

Basic Example

飛行場対空援助局（Hanamaki Radio）と交信

> **PIL:** Hanamaki Radio, JA 5807, leaving high station at 52.
>
> **AFIS:** JA 5807, roger. Report completing base turn.
>
> **PIL:** Report completing base turn, JA 5807.
>
>
> **PIL:** JA 5807, complete base turn.
>
> **AFIS:** JA 5807, roger, runway 20 runway is clear, wind 300 at 6.
>
> **PIL:** Runway 20 runway is clear, JA 5807.
>
>
> **AFIS:** JA 5807, taxi to spot H.
>
> **PIL:** Taxi to spot H, JA 5807.

Example 1

...MIYAKO (MQE) - MORIO - RJSI

Phraseology Example 1

　IFR をキャンセルする場合は，その旨を通報し，可能であれば必要に応じてその後のインテンションを通報する．「cancel IFR」を通報した後は，VFR なので VMC を維持しなければならない．なお，「cancel IFR」はパイロットが時間を遡ってすることはなく，また，管制機関（等）からキャンセルを要求されることは通常ない．

PIL:	Hanamaki Radio, JA 5809, departed MORIO.
AFIS:	JA 5809, make ILS Y runway 20 approach, wind 290 degrees at 9 knots, report localizer course and cancel IFR.
PIL:	JA 5809, report localizer, report cancel IFR.
PIL:	JA 5809, cancel IFR at 0404, continue simulated ILS Y runway 20 approach and touch and go.
AFIS:	JA 5809, cancel IFR at 0404, roger, (simulated ILS Y runway 20 approach, maintain VMC all the time), wind 280 at 8, report 10 miles on final.
PIL:	JA 5809, report 10 miles on final.
PIL:	JA 5809, 10 miles on final.
AFIS:	JA 5809, roger, runway 20 runway is clear, wind 310 degrees at 10 knots.

Phraseology Example 2

　IFR をキャンセルした後，Local Training を行う場合は，以下のようになる．

PIL:	Fukushima Radio, JA 5806, cancel IFR at 57, after touch and go, left turn to simulated ILS Z runway 01 approach.
AFIS:	JA 5806, roger, cancel IFR at 57, runway 01 runway is clear, wind 330 at 3.
PIL:	JA 5806, runway 01 runway is clear.
PIL:	JA 5806, after airborne, make left turn, report 11 miles south-southwest.
AFIS:	JA 5806, roger, copied.
PIL:	Fukushima Radio, JA 5806, departed 11 miles south-southwest, simulated ILS Z runway 01 approach and make 1 time touch and go.
AFIS:	JA 5806, report ABKMA.

Phraseology Example 3

　IFR をキャンセルする場合（模擬計器進入訓練やタッチアンドゴー等）は，雲に入ったり，VMC を維持できないような状況に陥らないように，気象状況を的確に判断しなければならない．IFR をキャンセルする場合は，雲からの距離を適正に確保して，VMC を維持した飛行を行う必要がある．

RDR:　JA 5801, descend and maintain 6,000.

PIL:　JA 5801, descend and maintain 6,000, request one time hold.

RDR:　JA 5801, one time hold, roger, hold east of Hakodate VOR one time, report completing hold.

PIL:　Hold east one time, report complete hold, JA 5801.

RDR:　JA 5801, descend and maintain 5,000, this time, report starting hold, report cancel IFR.

PIL:　JA 5801, descend and maintain 5,000, report starting hold, and report cancel IFR.

PIL:　Hakodate Radar, JA 5801, start hold, leaving 6,100 descend 5,000.

RDR:　JA 5801, roger, report complete hold.

PIL:　Report complete hold, JA 5801.

PIL:　Hakodate Radar, JA 5801, complete hold.

RDR:　JA 5801, cleared for VOR runway 12 approach, maintain 5,000 until further advised, report high station.

PIL:　JA 5801, cleared for VOR runway 12 approach, maintain 5,000 until further advised, report high station.

PIL:　JA 5801, leaving high station.

RDR:　JA 5801, you are number 2 landing traffic. Airbus 320 7 miles on final ILS approach.

PIL:　Roger, JA 5801, negative contact, we're in cloud.

RDR:　JA 5801, altitude restrictions cancelled, report starting base turn.

PIL:　JA 5801, altitude restrictions cancelled, report starting base turn.

PIL: Hakodate Radar, JA 5801, starting base turn, field in sight, cancel IFR.

RDR: JA 5801, this time, cleared for simulated VOR runway 12 approach, maintain VMC, squawk change 1520.

PIL: JA 5801, cleared for simulated VOR runway 12 approach, maintain VMC, squawk 1520, after touch and go, request simulated ILS Z runway 12 approach.

RDR: JA 5801, copied.

RDR: JA 5801, after completing touch and go, turn right heading 270, climb and maintain 3,500 for simulated ILS approach.

PIL: Roger, after touch and go, right turn heading 270, climb and maintain 3,500, JA 5801.

RDR: JA 5801, contact Tower 118.35.

PIL: 118.35, JA 5801.

PIL: Hakodate Tower, JA 5801, 8 miles final, touch and go.

TWR: JA 5801, Hakodate Tower, runway 12 cleared touch and go, wind 150 at 11.

PIL: Cleared touch and go, runway 12, JA 5801.

TWR: JA 5801, contact Radar.

PIL: Contact Radar, JA 5801.

PIL: Hakodate Radar, JA 5801, with you, leaving 1,900 for 3,500, heading 270.

RDR: JA 5801, Hakodate Radar, radar contact, maintain 3,500, continue present heading 270 vector to simulated ILS Z runway 12 final approach course.

PIL: JA 5801, maintain 3,500, heading 270.

　レーダー業務（radar service）とは，レーダーを使用して行う管制業務，飛行情報業務及び警急業務をいう．

　レーダーを使用して行う管制業務は，レーダー管制業務（radar control）と呼ばれ，レーダー間隔・レーダー監視・レーダー誘導がある．

　なお，飛行情報業務（flight information service）は，航空機の安全，かつ，円滑な運航に必要な情報を提供する業務であり，警急業務（alerting service）とは，捜索救難を必要とする航空機に関する業務である．

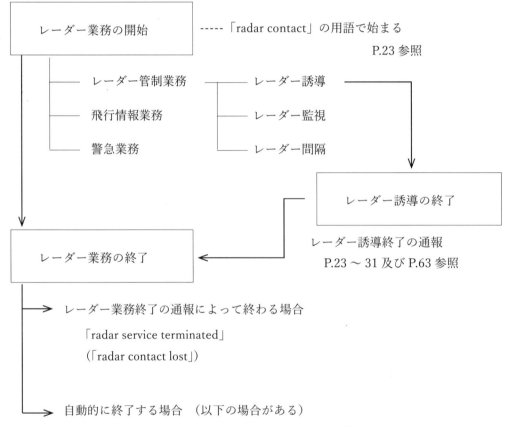

- レーダー業務の開始 ----- 「radar contact」の用語で始まる　P.23 参照
  - レーダー管制業務
  - 飛行情報業務
  - 警急業務
  - レーダー誘導
  - レーダー監視
  - レーダー間隔
- レーダー誘導の終了
  - レーダー誘導終了の通報　P.23 ～ 31 及び P.63 参照
- レーダー業務の終了
  - レーダー業務終了の通報によって終わる場合
    「radar service terminated」
    （「radar contact lost」）
  - 自動的に終了する場合　（以下の場合がある）
    1．航空機が計器飛行方式による飛行を取り下げた場合
    2．計器進入又は視認進入により着陸した場合
    3．管制区管制所が計器進入を許可し，飛行場管制所又は飛行場対空援助局と通信するよう指示した場合
    4．ターミナル管制所が計器進入を許可し，飛行場対空援助局と通信を設定するよう指示した場合
    5．PCA を飛行する VFR 到着機に対して，飛行場管制所と通信を設定するよう指示した場合

PART.3.

IFR Flight Scenario

PART.3. では, 以下の 2 つのフライトの ATC Communications を取り上げる.

UNIT.1.　Fukuoka airport (RJFF)　→　Miyazaki airport (RJFM)

　　　　　福岡空港　→　宮崎空港

UNIT.2.　Sendai airport (RJSS)　→　Niigata airport (RJSN)

　　　　　仙台空港　→　新潟空港

　一連の交信の流れに習熟するためのものなので説明等は付していない. 不明な点は PART.1. 及び PART.2. を参照されたい.

UNIT.1. RJFF → RJFM

Introduction

　UNIT.1. では，福岡空港から宮崎空港へ向かう航空機のシミュレーションを取り上げる．以下がフライトの主な内容である．

Aircraft Identification	Nippon Air 3615
Type of Aircraft	Bombardier Dash 8
Departure Aerodrome	Fukuoka airport
Level	FL 160
Route	KUE - ESKAP - KROMA - ENBEN - MZE
Destination Aerodrome	Miyazaki airport
SID	MORIO FOUR DEP
Transition	SAKURA TRANSITION
Instrument Approach Procedure	RNP X RWY 09 APP

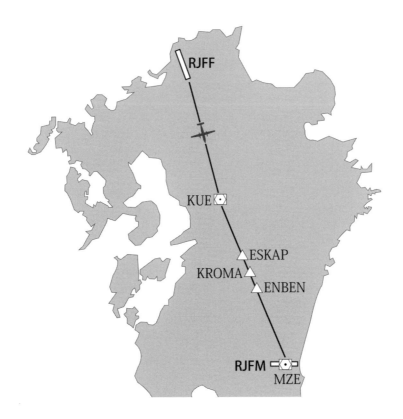

1. RJFF → RJFM

PIL: Fukuoka Delivery, Nippon Air 3615.

DEL: Nippon Air 3615, Delivery.

PIL: Nippon Air 3615, to Miyazaki airport, proposing FL 160, spot 17.

DEL: Nippon Air 3615 cleared to Miyazaki airport via Morio Four Departure, Sakura Transition, flight planned route, maintain FL 160, squawk 3405.

PIL: Nippon Air 3615 cleared to Miyazaki airport via Morio Four Departure, Sakura Transition, flight planned route, maintain FL 160, squawk 3405.

DEL: Nippon Air 3615, read back is correct, Ground frequency 121.7.

PIL: 121.7, Nippon Air 3615.

PIL: Fukuoka Ground, Nippon Air 3615, spot 17, with N, request taxi.

GND: Nippon Air 3615, Fukuoka Ground, runway 16, taxi via A to holding point E-2, cross GP hold line.

PIL: Nippon Air 3615, runway 16, taxi via A to holding point E-2, cross GP hold line.

GND: Nippon Air 3615, contact Tower 118.4.

PIL: 118.4, Nippon Air 3615.

PIL: Fukuoka Tower, Nippon Air 3615, ready.

TWR: Nippon Air 3615, Fukuoka Tower, wind 160 at 9, runway 16 at E-2, cleared for take-off.

PIL: Runway 16 at E-2, cleared for take-off, Nippon Air 3615.

TWR: Nippon Air 3615, contact Departure.

PIL: Contact Departure, Nippon Air 3615.

PIL: Fukuoka Departure, Nippon Air 3615, leaving 1,900 for FL 160.

DEP: Nippon Air 3615, Fukuoka Departure, radar contact.

PIL: Nippon Air 3615.

DEP: Nippon Air 3615, contact Kumamoto Approach 126.5.

PIL: 126.5, Nippon Air 3615.

PIL: Kumamoto Approach, Nippon Air 3615, climbing FL 160.

APP: Nippon Air 3615, Kumamoto Approach, roger.

PIL: Nippon Air 3615.

APP: Nippon Air 3615, descend to reach FL 150 by ESKAP.

PIL: Descend to reach FL 150 by ESKAP, Nippon Air 3615.

APP: Nippon Air 3615, contact Kagoshima Radar 121.4.

PIL: 121.4, Nippon Air 3615.

PIL: Kagoshima Radar, Nippon Air 3615, FL 160 assigned FL 150 by ESKAP, information H, request RNP X runway 09 approach via Melar Arrival.

RDR: Nippon Air 3615, Kagoshima Radar, roger, cleared via Melar Arrival, descend and maintain 12,000.

PIL: Cleared via Melar Arrival, descend and maintain 12,000, Nippon Air 3615.

RDR: Nippon Air 3615, descend via STAR to 5,200, cleared for RNP X runway 09 approach.

PIL: Descend via STAR to 5,200, cleared for RNP X runway 09 approach, Nippon Air 3615.

STANDARD ARRIVAL CHART - INSTRUMENT

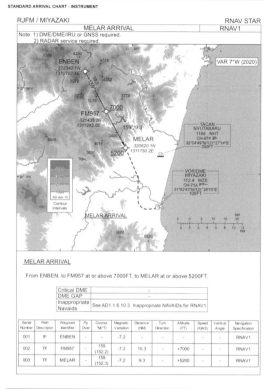

RJFM / MIYAZAKI	RNAV STAR
MELAR ARRIVAL	RNAV1

Note 1) DME/DME/IRU or GNSS required.
2) RADAR service required.

MELAR ARRIVAL

From ENBEN, to FM957 at or above 7000FT, to MELAR at or above 5200FT.

Critical DME	-
DME GAP	-
Inappropriate Navaids	See AD1.1.6.10.3. Inappropriate NAVAIDs for RNAV1

Serial Number	Path Descriptor	Waypoint Identifier	Fly Over	Course °M(°T)	Magnetic Variation	Distance (NM)	Turn Direction	Altitude (FT)	Speed (KIAS)	Vertical Angle	Navigation Specification
001	IF	ENBEN	-	-	-7.2	-	-	-	-	-	RNAV1
002	TF	FM957	-	159 (152.2)	-7.2	10.3	-	+7000	-	-	RNAV1
003	TF	MELAR	-	159 (152.3)	-7.2	9.3	-	+5200	-	-	RNAV1

INSTRUMENT APPROACH CHART

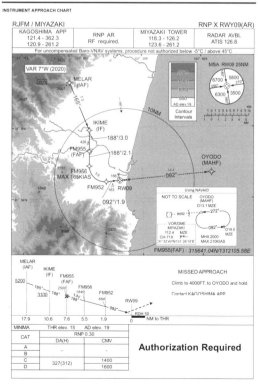

RJFM / MIYAZAKI			RNP X RWY09(AR)
KAGOSHIMA APP 121.4 - 362.3 120.9 - 261.2	RNP AR RF required.	MIYAZAKI TOWER 118.3 - 126.2 123.6 - 261.2	RADAR AVBL. ATIS 126.8

For uncompensated Baro-VNAV systems, procedure not authorized below -5°C / above 45°C

FM955(FAF) : 315641.04N/1312105.58E

MISSED APPROACH

Climb to 4000FT, to OYODO and hold.

Contact KAGOSHIMA APP.

Authorization Required

MINIMA	THR elev. 15	AD elev. 19
CAT	RNP 0.30	
	DA(H)	CMV
A	-	-
B		
C	327(312)	1400
D		1600

RDR: Nippon Air 3615, contact Miyazaki Tower 118.3.

PIL: Tower 118.3, Nippon Air 3615.

PIL: Miyazaki Tower, Nippon Air 3615, departed IKIME.

TWR: Nippon Air 3615, Miyazaki Tower, continue approach, wind 060 at 9.

PIL: Continue approach, Nippon Air 3615.

TWR: Nippon Air 3615, runway 09 cleared to land, wind 050 at 8.

PIL: Runway 09 cleared to land, Nippon Air 3615.

TWR: Nippon Air 3615, contact Ground 121.9.

PIL: Ground 121.9, Nippon Air 3615.

PIL: Miyazaki Ground, Nippon Air 3615, pick up S-6, spot 6.

GND: Nippon Air 3615, Miyazaki Ground, taxi to spot 6.

PIL: Taxi to spot 6, Nippon Air 3615.

UNIT.2. RJSS → RJSN

Introduction

UNIT.2. では，仙台空港から新潟空港へ向かう航空機のシミュレーションを取り上げる．以下がフライトの主な内容である．

Aircraft Identification	JA 5810
Type of Aircraft	Beechcraft G58 Baron
Departure Aerodrome	Sendai airport
Level	10,000 feet
Route	SDE - R-217 - GTC
Destination Aerodrome	Niigata airport
SID	SENDAI REVERSAL SIX DEP
Instrument Approach Procedure	VOR RWY 10 APP

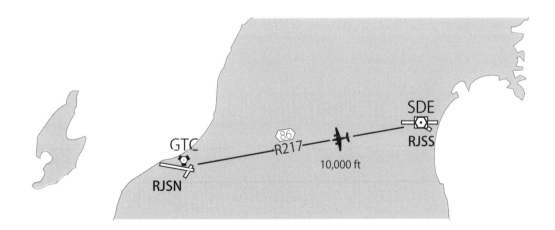

2. RJSS → RJSN

GND: JA 5810 cleared to Niigata airport via Sendai Reversal Six Departure, flight planned route, maintain 10,000, squawk 2374.

PIL: JA 5810 cleared to Niigata airport via Sendai Reversal Six Departure, flight planned route, maintain 10,000, squawk 2374.

GND: JA 5810, read back is correct.

PIL: Sendai Ground, JA 5810 at R area, request taxi to B-3 intersection departure.

GND: JA 5810, B-3 intersection approved, taxi to holding point A-1 runway 12.

PIL: B-3 intersection approved, taxi to holding point A-1 runway 12, JA 5810.

GND: JA 5810, cross runway 12, taxi to holding point B-3 runway 09.

PIL: JA 5810, cross runway 12, taxi to holding point B-3 runway 09.

GND: JA 5810, contact Tower 118.7.

PIL: Contact Tower 118.7, JA 5810.

PIL: Sendai Tower, JA 5810, taxi to B-3, ready.

TWR: JA 5810, Sendai Tower, hold short of runway 09 at B-3, you are number 2 departure.

PIL: Hold short of runway 09 at B-3, number 2, JA 5810.

TWR: JA 5810, runway 09 at B-3, line up and wait.

PIL: Runway 09 at B-3, line up and wait, JA 5810.

TWR: JA 5810, wind 120 at 12, runway 09 cleared for take-off.

PIL: Runway 09 cleared for take-off, JA 5810.

TWR: JA 5810, contact Sendai Approach 120.4.

PIL: Contact Sendai Approach 120.4, JA 5810.

PIL: Sendai Approach, JA 5810, leaving 1,300 assigned 10,000.

APP: JA 5810, Sendai Approach, radar contact, fly heading 120 vector to Sendai VOR, climb and maintain 10,000.

PIL: Fly heading 120, climb and maintain 10,000, JA 5810.

APP: JA 5810, turn left heading 360.

PIL: Turn left heading 360, JA 5810.

APP: JA 5810, resume own navigation direct Sendai VOR.

PIL: Resume own navigation direct Sendai VOR, JA 5810.

APP: JA 5810, contact Tokyo Control 118.9.

PIL: Contact Tokyo Control 118.9, JA 5810.

PIL: Tokyo Control, JA 5810, maintain 10,000.

ACC: JA 5810, Tokyo Control, area QNH 3007.

PIL: Area QNH 3007, JA 5810.

ACC: JA 5810, contact Tokyo Control 132.3.

PIL: Contact Tokyo Control 132.3, JA 5810.

PIL: Tokyo Control, JA 5810, maintain 10,000.

ACC: JA 5810, Tokyo Control, area QNH 2990.

PIL: Area QNH 2990, JA 5810.

ACC: JA 5810, contact Niigata Approach 121.4.

PIL: Contact Niigata Approach 121.4, JA 5810.

PIL: Niigata Approach, JA 5810, maintain 10,000, information K.

APP: JA 5810, Niigata Approach, information L, QNH 2990.

PIL: JA 5810, information L, QNH 2990, request direct Niigata VORTAC and VOR runway 10 approach from high station.

APP: JA 5810, recleared direct Niigata VORTAC, descend at pilot's discretion maintain 8,000.

PIL: Recleared direct Niigata VORTAC, descend at pilot's discretion maintain 8,000, JA 5810.

APP: JA 5810, descend at pilot's discretion maintain 5,000.

PIL: Descend at pilot's discretion maintain 5,000, JA 5810.

APP: JA 5810, descend and maintain 4,000, upon Niigata VORTAC, cleared for VOR runway 10 approach, report high station.

PIL: Descend and maintain 4,000, upon Niigata VORTAC, cleared for VOR runway 10 approach, report high station, JA 5810.

PIL: Niigata Approach, JA 5810, leaving high station.

APP: JA 5810, report starting base turn.

PIL: Report starting base turn, JA 5810.

PIL: JA 5810, starting base turn.

APP: JA 5810, contact Niigata Tower 118.0.

PIL: Contact Niigata Tower 118.0, JA 5810.

PIL: Niigata Tower, JA 5810, on base turn, full stop.

TWR: JA 5810, Niigata Tower, runway 10 cleared to land, wind 120 at 6.

PIL: Runway 10 cleared to land, JA 5810.

TWR: JA 5810, turn right B-2, request intention.

PIL: Turn right B-2, request taxi to B-4, JA 5810.

TWR: JA 5810, runway 10, taxi to holding point B-4.

PIL: Runway 10, taxi to holding point B-4, JA 5810.

PIL: Niigata Ground, JA 5810, at B-4, request IFR clearance to Sendai airport, proposing 11,000.

Niigata airport information K, 0600, VOR runway 10 approach, using runway 10 and 22, wind 150 degrees at 14 knots, visibility 40 km, few 4,000 feet cumulus, broken height unknown, temperature 23, dew point 9, QNH 2990 inches, advise you have information K.

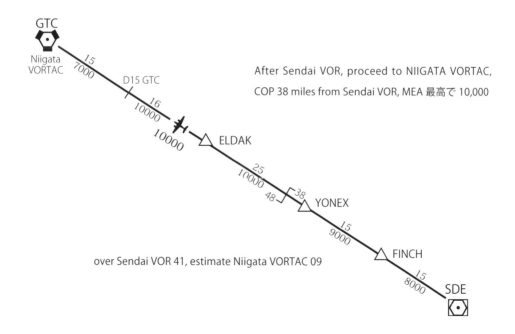

After Sendai VOR, proceed to NIIGATA VORTAC, COP 38 miles from Sendai VOR, MEA 最高で 10,000

over Sendai VOR 41, estimate Niigata VORTAC 09

PART.4.

Miscellaneous Expressions

　PART.4. では，PART.1. 及び PART.2. で扱わなかった事項を中心に取扱う．

　初期課程の訓練生は，ここで扱う場面に実際に遭遇する，又は同種の交信を行う可能性は比較的少ないと思われるが，他の航空機の交信・行動の把握は非常に重要であるという観点から，学習者の知識の構築に役立てるため取扱っている．

　更に，ICAO LEVEL 4 関連でも，航空英語能力証明試験の学科試験のリスニング問題，及び，実地試験のロールプレイングの際にはこれらの知識も必要となってくるであろう．

UNIT.1. Emergency

Words & Phrases

ditching	minimum fuel (*1)
不時着水	ミニマム　フューエル
confirm you are squawking 7500	understand your situation, 7500 observed
7500 を発信していますか	状況は分かりました，7500 確認しました
squawk Mayday (code 7700)	
緊急コード (7700) を送って下さい	
radar contact, if feasible, squawk ~	
レーダーコンタクト，可能ならば～を送って下さい	
recycle ~	medical emergency
～を再設定して下さい	医療に関する緊急事態
remaining fuel	souls on board
残存燃料	搭乗人員
stop transmitting, MAYDAY	cancel distress
通信停止，メーデー	遭難を取り消します
distress traffic ended	interception
遭難終了	要撃

＊ (*1) パイロットは，着陸を計画する飛行場の選択肢が一つの飛行場に限定され，着陸を決心した飛行場に安全に着陸するために必要な残存燃料が遅延を受け入れられない状態に達したときには，管制機関等に対して「minimum fuel」の状態であることを通報しなければならない．「minimum fuel」の通報は，優先的取扱いを要求するものではなく，遅延が生じれば緊急状態になる潜在性を意味している．燃料欠乏の状態が緊急状態であると宣言（Mayday Mayday Mayday fuel / Mayday fuel）すれば，管制上優先的取扱いが行われる．

Introduction

　パイロットは，緊迫した状態が発生したら，緊急事態を宣言して，管制上優先的取扱いの支援を要請する．重大な事態に直面したパイロットは，

　　1．可能であれば VMC を維持して上昇

　　2．管制機関と通信を設定

　　3．通信設定できない場合はトランスポンダーを 7600 又は 7700 にセット

を行うのが望ましい．周波数はそれまで使用中の周波数によって行うが，緊急用周波数 121.5 又は 243.0 を使用しても構わない．なお，管制機関より使用周波数を指定された場合はその周波数を使用する．

　遭難通信は「Mayday」，緊急通信は「Pan Pan」の信号で開始しなければならない．日本では，「Emergency」の宣言によって，管制上の優先措置が取られるが，国際的には「Mayday」や「Pan Pan」の用語を使用することとなっている．「Emergency」は緊急信号ではない．

・Distress --- a condition of being threatened by serious and / or imminent danger and of requiring immediate assistance.

「重大か且つ / 又は急迫した危険にさらされており，且つ即時の援助を必要とする状態」

・Urgency --- a condition concerning the safety of an aircraft or other vehicle, or of some person on board or within sight, but which does not require immediate assistance.

「航空機もしくは他の輸送媒体の安全，又は，機上もしくは視界内の人の安全に関する状態であって，即時の援助を必要としない状態」

　なお，上記，「遭難」（Distress；Mayday）と「緊急」（Urgency；Pan Pan）の目安としては，管制の援助を受ける等して，飛行場等の着陸に適した場所に到達できる見込みがあり，消火・救難の必要性がない状態までを緊急とし，不時着の可能性が高い状況又は飛行場へ着陸後，消火・救難が必要な場合を遭難と判断する．

Phraseology Example 1

　機材の故障，残存燃料の不足等，危険を感じ直ちに管制官等の援助を必要とする場合は，遅滞なく「Mayday」を通報するべきである．口頭によるほか，トランスポンダーのコードでもできる．

PIL:	Mayday, Mayday, Mayday, Tokyo Approach, Nippon Air 713, need to return to Narita immediately. Fire in left engine.
APP:	Nippon Air 713, roger Mayday. Turn right heading 070 vector to ILS runway 34L final approach course.
PIL:	Heading 070, Nippon Air 713.
APP:	Nippon Air 713, when ready, report remaining fuel and souls on board and squawk 7700.
PIL:	We have 1.5 hours endurance of fuel, and 224 souls on board, transponder to 7700, Nippon Air 713.

　遭難状態を管制機関が了承した場合，特に指示されない限りは，管制機関より指定されたトランスポンダーコードを作動させておく．

　もし，通信設定ができない場合はトランスポンダーを 7600 又は 7700 にセットする．

　通信設定ができた場合は，航空機は管制機関の指示に従った飛行を行う．

　飛行中に管制機関との交信ができなくなった航空機は，一般的には緊急状態と認識され，管制上優先的取扱いが取られる．しかし，緊急状態を宣言しない限りは，トランスポンダーを 7600 にセットし航空法施行規則第 206 条に沿った飛行を行わなければならない．

　また，現在位置が不確実の場合，不時着（水）等，差し迫った状態に陥ったとき，ELT（航空機の遭難や墜落等の際に，その地点を探知させるための信号を送信する装置）を装備している場合は，それを作動させるべきである．

Phraseology Example 2

「Mayday」及び「Pan Pan」の定義は P.139 の通りであるが，このどちらを選ぶかは，通常，「Distress」及び「Urgency」の定義により機長が判断する．また，「Pan Pan」を宣言したものの，その後状況が悪化した場合，「Mayday」を宣言することもありうる．

PIL: Pan-Pan, Pan-Pan, Pan-Pan, Nippon Air 810, request descend to FL 200, we received an indication of the loss of cabin pressure.

ACC: Nippon Air 810, roger Pan. Descend and maintain FL 200.

PIL: Descend FL 200, Nippon Air 810.

PIL: Mayday, Mayday, Mayday, Nippon Air 810, we request an emergency descent to resume normal pressure of the aircraft, we have a pressure issue.

ACC: Nippon Air 810, roger Mayday. Descend and maintain 10,000.

PIL: Descending 10,000, Nippon Air 810. Some of the passengers are feeling unwell. I'll get back to you.

✈ POINT:　管制上優先的取扱い

1．航空機が「メーデー」又は「パン　パン」を通報した場合

2．航空機が残存燃料について緊急状態である旨を通報した場合（「Mayday Fuel」又は「Mayday Mayday Mayday Fuel」と通報された場合は遭難の段階として取扱う）

3．航空機が発動機の故障等により緊急状態にある旨を通報した場合

4．二次レーダーコード 7700 の表示をレーダー画面上に観察した場合

5．その他，航空機が明らかに緊急状態にあって優先的に取扱う必要があると認められる場合

6．急病人若しくは重病人又は移植臓器を搬送している航空機又は臓器の移植を目的として運航している航空機が，優先的取扱いを要求した場合又は優先的に取扱う必要があると認められる場合

7．災害派遣又は人命財産の保護のために緊急に出動する航空機が，優先的取扱いを要求した場合又は優先的に取扱う必要があると認められる場合

8．航空機若しくは運航者から不法妨害を受けている旨通報された場合又はそのおそれがあると認められる場合

9．航空機が ADS-C の緊急機能を作動させた場合

10．航空機が CPDLC 経由で緊急状態を示す旨のメッセージを送信した場合

11．航空機が火山灰雲に入った旨通報した場合

Phraseology Example 3

「Distress」は「Mayday」を前置し，「Urgency」は「Pan Pan」を前置するとなっており，「Emergency」は緊急信号ではないが，日本では「Emergency」の宣言によっても管制上の優先措置が受けられる．

PIL: Runway 34R at C4B, line up and wait, Nippon Air 155.

TWR: Nippon Air 155, wind 330 at 12, runway 34R cleared for take-off.

PIL: Nippon Air 155, runway 34R cleared for take-off.

TWR: Nippon Air 155, contact Departure.

PIL: Contact Departure, Nippon Air 155.

PIL: Tokyo Departure, Nippon Air 155, leaving 2,000 climbing FL 150.

DEP: Nippon Air 155, Tokyo Departure, radar contact.

DEP: Nippon Air 155, fly heading 090 vector to Sekiyado, maintain FL 150.

PIL: Nippon Air 155, we have one engine failure, number 2 right engine failure, declare emergency.

DEP: Nippon Air 155, confirm declare emergency?

PIL: Affirm, Nippon Air 155.

DEP: Roger, copied, maintain 4,000.

PIL: Maintain 4,000, Nippon Air 155.

DEP: Nippon Air 155, do you accept ILS approach?

PIL: Nippon Air 155, request ILS Z runway 34R approach.

DEP: Nippon Air 155, roger, vector to final, runway 34R.

PIL: Roger, Nippon Air 155.

DEP: Nippon Air 155, turn right heading 160, maintain 4,000.

PIL: 160, 4,000, Nippon Air 155.

DEP: Nippon Air 155, contact Tokyo Approach 119.7.

UNIT.2. Information (Pilot to ATC)

Words & Phrases

pilot reports

　　パイロットレポート，機上気象報告

request flight conditions

　　飛行状況を通報して下さい

clear air turbulence

　　晴天乱気流

Introduction

　パイロットは飛行中，運航の妨げとなる気象現象に遭遇した場合は管制機関（等）に通報する（なお，この気象報告は PIREP と呼ばれる）．航空機が管制機関（等）に対して提供した気象情報等は，関係空域を飛行する航空機にその内容が通報される．

　通常，パイロットは，その通報に際し，

　1．位置

　2．（飛行）高度

　3．観測時刻

　4．航空機の型式

を含めた飛行状況を通報することが望ましい．

　なお，管制機関が航空機に気象に関する情報を要求する場合は「request flight conditions」の用語が用いられる．

　パイロットは，通報した状況が他の航空機に対しても有益であることから，状況をパイロットの目線（何が起こったのか，それからどうしたのか，どういうことが必要だったのか等）を念頭において通報することが望ましい．

Phraseology Example 1

運航に影響を及ぼすような気象状況を発見・遭遇したら，可能な限り管制機関（等）に通報する．

晴天乱気流の場合

PIL:	Tokyo Control, Nippon Air 810, PIREP.
ACC:	Nippon Air 810, Tokyo Control, go ahead.
PIL:	Nippon Air 810, experience moderate clear air turbulence over DAIGO, FL 280, 2 minutes ago, Boeing 737.
ACC:	Nippon Air 810, roger, do you request a change in altitude?
PIL:	Negative, we seem to be getting into some smoother air now.
ACC:	Nippon Air 810, roger.

タービュランスの場合

PIL:	Tokyo Control, Nippon Air 810, we are encountering light plus turbulence about 20 miles east of BOBOT at FL 350. Request descent to FL 250.
ACC:	Nippon Air 810, Tokyo Control, descend and maintain FL 250. Report speed.

PIL:	Climb and maintain FL 240, recleared direct TOHME, Nippon Air 710.
PIL:	Tokyo Control, Nippon Air 710, pilot report.
ACC:	Nippon Air 710, go ahead.
PIL:	Nippon Air 710, around TAURA, altitude between 7,000 and FL 140, moderate turbulence, Boeing 767.
ACC:	Nippon Air 710, just confirmation, position, around TAURA, 7,000 to FL 140, is that correct?
PIL:	Affirm, Nippon Air 710.

雷雨の場合

PIL:	Tokyo Control, Nippon Air 810.
ACC:	Nippon Air 810, go ahead.
PIL:	Nippon Air 810, observed CB with lightning 40 miles ahead, cloud top appears FL 390 or higher, request deviation to the southeast about 20 miles.

着氷の場合

PIL:	Fukuoka Control, JA 5802, encountering severe icing between Okayama and Takamatsu, time 0800 to 0810, altitude 9,000 in cloud. Beechcraft Baron G58. Request descend to 7,000.

気象状況等に関する用語

SNOW

Dry
Wet
Dry snow
Wet snow
Slush
Compacted snow
Ice
Wet ice
Dry snow on top of compacted snow
Wet snow on top of compacted snow
Water on top of compacted snow
Dry snow on top of ice
Wet snow on top of ice

ICING

Trace
Light
Moderate
Severe

BRAKING ACTION

Good
Medium to Good
Medium
Medium to Poor
Poor
Less than Poor

TURBULENCE

Light
Moderate
Severe

　なお，タービュランスに関して，「light」の中でも特に強めのものは「light plus」，弱めのものは「light minus」と表現されることが多い．なお，揺れない状態は「smooth」と表現される．

低高度ウィンドシアーの場合

> PIL: Miyazaki Tower, Nippon Air 619, encountered wind shear on final, gained 25 knots between 600 and 400 followed by loss of 40 knots.
>
> TWR: Nippon Air 619, copied, contact Ground 121.7.

ウィンドシアーを通報する際に，特段決まった用語・フォーマット等はないが，

1．速度と高度の損失　　　2．そのときの措置　　　3．風向風速

等を通報することが望ましい．

（以下，用例）

PIL: Wind fairly gradual decrease, 50 knots indicates at 2,000 and mostly 30 knots on the approach. No wind shear on final.

PIL: Fluctuating. / Just a lot of rolling. / Really rough.

PIL: Below 2,000 plus minus 15 knots.

PIL: At 2,000, 320 degrees 50 knots.

PIL: Maximum thrust required.

　ウィンドシアー回避（wind shear escape）による管制指示からの逸脱について，パイロットからの通知がなければ管制機関はウィンドシアー回避により管制指示から逸脱していることを知り得ないため，福岡 FIR 内を飛行するすべてのウィンドシアー警報システム装備機のパイロットは，ウィンドシアー回避のため管制指示に従うことが困難な場合は，業務量が許す範囲において，可能な限り速やかに管制機関に通知する．

　ウィンドシアー回避の関連用語には，以下のものがある．

１．ウィンドシアー回避により管制指示からの逸脱を開始したとき

　PIL: Nippon Air 810, wind shear escape.

２．ウィンドシアー回避終了後，管制指示へ復帰するとき

　PIL: Nippon Air 810, wind shear escape complete, resuming last assigned heading.

　PIL: Nippon Air 810, wind shear escape complete, resuming Kizna Two Departure, maintaining 1,000.

３．ウィンドシアー回避終了後，新たな管制指示を要求するとき

　PIL: Nippon Air 810, wind shear escape complete, request ～

　PIL: Nippon Air 810, wind shear escape complete, request further instructions.

４．ウィンドシアー回避のため，受領した管制指示に従うことができないとき

　PIL: Nippon Air 810, unable, wind shear escape.

　なお，ウィンドシアー回避により管制指示から逸脱を行う場合，パイロットは航空法第 96 条第 1 項の航空交通の指示の違反には問われない．

147

UNIT.3. Information (ATC to Pilot)

Words & Phrases

fuel dump	fuel jettison
燃料投棄	燃料投棄
emergency descent	out of service, not available, unserviceable
緊急降下	運用停止中です
runway close	disabled aircraft
滑走路閉鎖	航行不能航空機
bomb disposal	earthquake
不発弾処理	地震
seismic	seismic intensity
地震の	震度
volcanic ash	(volcanic) ash cloud
火山灰	火山灰雲，噴煙
eruption, explosion	Tower observation
噴火，爆発	タワー観察
launch	yellow sand (dust)
（ロケット・ミサイル等の）発射	黄砂

Introduction

　航空保安施設の機能障害，運航の妨げとなる可能性がある気象情報等，航空機の安全運航上必要と認められる場合は，管制機関（等）より航空機に対してその情報が提供される．

Phraseology Example 1

燃料投棄の実施と終了に関する内容は，周辺の航空機に以下のように通報される．

> PIL: Tokyo Departure, Nippon Air 713, lost one engine, request fuel dump about 45 minutes then return to Narita.
>
> DEP: Nippon Air 713, Tokyo Departure, radar contact, fly heading 120 vector for fuel dump, maintain 7,000, you may commence dumping any time, report completion.
>
> PIL: Nippon Air 713, heading 120, maintain 7,000, report completion.
>
> DEP: All stations, Tokyo Departure, fuel dumping in progress over VENUS at 7,000 by Boeing 747.
>
> DEP: All stations, Tokyo Departure, fuel dumping over VENUS terminated.

なお，燃料投棄の必要が生じた航空機は，管制機関（等）と燃料投棄を行う経路・地点・高度等を調整し指示を受ける．その際，機種により投棄中無線機の発信ができないものがあるため，通信途絶前にパイロットは所要時間等必要な事項を通報することが望ましい．

通常，燃料投棄は，

1．原則として 6,000 ft 以上の高度

2．水平距離 10 マイル，上方 1,000 ft 下方 3,000 ft 以上の間隔

のもと行われる．

Phraseology Example 2

航空機が緊急降下している旨の通報を受けた管制機関は，可能な限り速やかに周辺を飛行する航空機に対し，当該航空機の緊急降下について情報を提供することがある．

> ACC: All stations, Tokyo Control, emergency descent in progress, 10 miles south of PABBA, Boeing 767 from FL 350 to 13,000, eastbound.

Phraseology Example 3

　航空機の運航に影響を及ぼすおそれがある場合であって（震度4以上 (*1)），緊急地震速報が出たとき，又は当該飛行場で地震が観測されたとき，管制機関から航空機に対して口頭でその旨提供される場合がある（多機能型地震計が設置された空港 (*2) で提供される）．

> PIL: Nippon Air 652, runway 22 cleared to land.
>
> TWR: Nippon Air 652, go around due to earthquake, seismic intensity 4.
> PIL: Go around, Nippon Air 652.
>
> TWR: All stations, earthquake information, seismic intensity 4, runway inspection will be made.

　着陸機に関しては上記のように復行が指示され，離陸機に対しては，「cancel take-off clearance due to earthquake」のように離陸許可の取り消しがなされる場合がある．

　なお，緊急地震速報が発出された場合は，

「all stations, earthquake early warning was issued, use caution」のようにその情報が提供される場合がある．

＊（*1）震度3以下であっても，情報が提供される場合がある．
＊（*2）2023年2月現在，新千歳・仙台・成田・羽田・中部・大阪・関空・福岡・鹿児島・那覇・新潟・広島・高松に設置されている．

Phraseology Example 4

　タービュランスに関する情報は以下のように提供される．

> ACC: Nippon Air 810, Tokyo Control.
> PIL: Nippon Air 810, go ahead.
> ACC: Nippon Air 810, information. Moderate turbulence was reported by Boeing 777 between FL 350 and FL 330, flying 10 minutes ahead of you.
> PIL: Thank you for your information, Nippon Air 810. How about the condition at FL 310.
> ACC: Nippon Air 810, FL 310 is smooth air now.
> PIL: Nippon Air 810, roger, request descent to FL 310.

Phraseology Example 5

火山灰に関する情報は以下のように提供される.

ACC: Nippon Air 810, Tokyo Control.

PIL: Nippon Air 810, go ahead.

ACC: Nippon Air 810, volcanic ash cloud is reported over Miyakejima area, almost 50NM ahead of you. The top of the ash cloud was observed higher than FL 410, spreading southwest. Can you observe that cloud?

PIL: Nippon Air 810, affirm. We can see very dense ash cloud ahead. Request heading 350 for another 20 miles to avoid the ash cloud.

噴火状況が管制機関（等）より一斉送信される場合は,「all stations, Kagoshima Tower, Mt. Sakurajima erupted at 0130, top of ash cloud 9,000, moving northeast」等のように行われる.

パイロットが火山灰を発見した場合は, 速やかにその状況を管制機関（等）に通報すべきである. また, 火山灰に遭遇した場合は速やかに火山灰雲から脱出するべきである. なお, 噴煙は「wispy」（薄い）,「moderate dense」（濃い）,「very dense」（極く濃い）と表現される.

Phraseology Example 6

ウィンドシアー又はマイクロバーストに関する情報も提供される. ATIS により放送される場合は「wind shear advisories in effect」「microburst advisories in effect」の用語が使用される. また, パイロットレポートの情報も提供される場合がある.

PIL: Narita Tower, Nippon Air 714, 6 miles on final, runway 34L.

TWR: Nippon Air 714, Narita Tower, runway 34L cleared to land, wind 320 at 29, maximum 40, minimum 20. Also, PIREP wind shear below 2,000, plus minus 15 knots, reported Boeing 767.

PIL: Thank you, Nippon Air 714, cleared to land, runway 34L.

TWR: Narita Tower broadcast, PIREP, wind shear on final, runway 34L, below 2,000, plus minus 15 knots. Surface winds 320 at 23 knots, maximum 34, minimum 15. Out.

マイクロバーストの場合も，同様である．

TWR:	All stations, microburst alert, runway 34R arrival, 44 knots loss, 2 miles final, use caution.
PIL:	Tokyo Tower, Nippon Air 154, approaching CACAO, gate 56.
TWR:	Nippon Air 154, Tokyo Tower, runway 34R continue approach, wind 330 at 12.
PIL:	Continue approach, runway 34R, Nippon Air 154.
PIL:	Tokyo Tower, Nippon Air 154, go around.
TWR:	Nippon Air 154, roger, continue present heading, maintain 4,000.
PIL:	Nippon Air 154, continue present heading, maintain 4,000.
TWR:	Nippon Air 154, execute missed approach course, maintain 4,000.
PIL:	Nippon Air 154, execute missed approach course, maintain 4,000.
TWR:	Nippon Air 154, contact Departure 126.0.

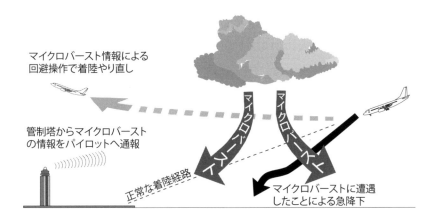

マイクロバースト情報による
回避操作で着陸やり直し

管制塔からマイクロバースト
の情報をパイロットへ通報

正常な着陸経路

マイクロバーストに遭遇
したことによる急降下

UNIT.4. Miscellaneous

Phraseology Example 1

何らかの理由により，目的地を変更する場合は，以下のようになる．

ACC:	Nippon Air 810, turn left heading 200.
PIL:	Nippon Air 810, turn left heading 200, request destination change to Narita airport due to smoke in the cabin.
ACC:	Nippon Air 810, roger, destination Narita copied, stand by.
PIL:	Nippon Air 810, stand by.
ACC:	Nippon Air 810, descend at pilot's discretion maintain FL 230.
PIL:	Nippon Air 810, descend at pilot's discretion maintain FL 230.
ACC:	Nippon Air 810, turn left heading 150.
PIL:	Nippon Air 810, turn left heading 150.
ACC:	Nippon Air 810, revised clearance.
PIL:	Go ahead.
ACC:	Nippon Air 810 cleared to Narita airport via radar vector to GUPER, Y-81, RUTAS then direct, maintain FL 230, continue heading 150.

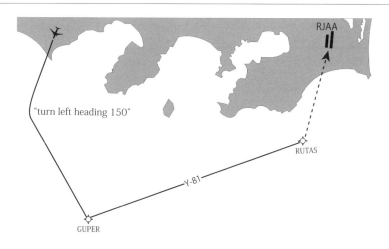

Phraseology Example 2

　管制指示と異なる RA が発生した場合，パイロットは，可能な限り，速やかに管制機関にその旨を通報するべきである．

PIL:　　Naha Departure, Nippon Air 810, 1,000.

DEP:　　Nippon Air 810, Naha Departure, radar contact, maintain 1,000.

PIL:　　Maintain 1,000, Nippon Air 810.

DEP:　　Nippon Air 810, traffic, 12 o'clock, 5 miles northeastbound, fighters inbound to Kadena 3,000.

PIL:　　Nippon Air 810, negative contact.

PIL:　　Nippon Air 810, TCAS RA.

DEP:　　Nippon Air 810, roger, do you have the F-fifteens in sight? Maintain 1,000, traffic fighters of two F-fifteens leaving 2,000.

PIL:　　Nippon Air 810, maintain 1,000.

DEP:　　Nippon Air 810, clear of traffic, climb and maintain FL 260.

　TCAS（Traffic alert and Collision Avoidance System）の関連用語には以下がある．

１．TCAS の通報を行うとき

　PIL: Nippon Air 810, TCAS RA.

２．TCAS RA の発生により ATC の指示に従えない場合

　PIL: Nippon Air 810, unable, TCAS RA.

３．TCAS が [clear of conflict] を表示した場合

　　３－１．assigned clearance（例えば，9,000 ft）に戻ろうとする場合

　PIL: Nippon Air 810, clear of conflict, returning to 9,000.

　　３－２．assigned clearance に戻った時点で通報する場合

　PIL: Nippon Air 810, clear of conflict, 9,000 resumed.

　TCAS RA と ATC（管制機関）の衝突回避指示に違いがあった場合，パイロットは RA に従わなければならない．なお，RA の指示による指示高度からの逸脱は，航空法第 96 条第 1 項の航空交通の指示の違反には問われない．

Phraseology Example 3

平行滑走路のある飛行場では，両方の滑走路を使用して，

1．平行 ILS 進入・平行 ILS/PAR 進入：

規定のレーダーセパレーションを適用する

2．同時平行 ILS 進入・同時平行 ILS/PAR 進入：

NTZ（不可侵区域）を設定，レーダーセパレーションを適用しない

が行われる場合がある．

それぞれ，進入に先立ち，以下の用語によって情報が提供される（ATIS 等に含まれている場合は省略される）．

APP: Parallel ILS approaches to runway 34L and right are in progress.

APP: Simultaneous parallel ILS approaches to runway 34L and R are in progress.

ATIS によって，情報が提供される場合は，以下のように行われる．

Tokyo international airport, information Q, 0000, ILS Z runway 34L approach and ILS Z runway 34R approach, landing runway 34L and 34R, departure runway 05 and 34R, Departure frequency 126.0 from runway 05, 120.8 from runway 34R, simultaneous parallel ILS approaches to runway 34L and right in progress, wind 340 degrees 12 knots, visibility 15 km, few 2,000 feet cumulus, broken 2,500 feet stratocumulus, temperature 10, dew point 6, QNH 3024 inches, advise you have information Q.

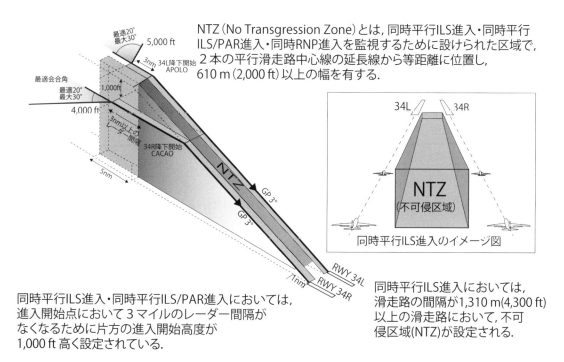

NTZ（No Transgression Zone）とは，同時平行ILS進入・同時平行ILS/PAR進入・同時RNP進入を監視するために設けられた区域で，2本の平行滑走路中心線の延長線から等距離に位置し，610 m（2,000 ft）以上の幅を有する.

同時平行ILS進入のイメージ図

同時平行ILS進入・同時平行ILS/PAR進入においては，進入開始点において3マイルのレーダー間隔がなくなるために片方の進入開始高度が1,000 ft 高く設定されている.

同時平行ILS進入においては，滑走路の間隔が1,310 m(4,300 ft)以上の滑走路において，不可侵区域(NTZ)が設定される.

155

Phraseology Example 4

着陸誘導管制所（GCA）は管制官が IFR 機に対して，レーダーによって着陸のための誘導を行うもので，訓練時や，悪天時や ILS を備えていない航空機，戦闘機等に対して行われる．以下の２つがある．

1．精測レーダー進入　（PAR Approach：精密進入）

→　精測レーダーにより，最終進入コースの指示と，グライドパスからの偏位情報の提供を行う．

2．捜索レーダー進入　（Surveillance Approach：非精密進入）

→　空港監視レーダー（ASR）により，飛行高度の指示，降下パスについての助言を行う．

PAR Approach においては，管制官は航空機に対して，一方的かつ連続的に指示を行い，航空機を誘導限界点まで誘導する．なお，ILS の CAT I に相当する．

最終進入開始前（パターン席：ASR のレーダー席との交信）

1．誘導形式・滑走路・誘導限界が通報される

PIL:　　Naha GCA, Nippon Air 810.

GCA:　　Nippon Air 810, Naha GCA, this will be a PAR approach to runway 36, guidance limit 209 feet.（*1）

2．飛行場の気象情報（ATIS に含まれる場合は省略）

3．通信連絡途絶時の飛行方法の指示（IMC の場合，又は IMC の可能性がある場合）

GCA:　　If no transmissions are received for 1 minute in the pattern or 5 seconds on final approach, attempt contact Tower 118.1 and proceed with VOR runway 36 approach.

4．進入復行方式（省略される場合もある）

GCA:　　Your missed approach procedure is the same as the VOR runway 36 missed approach procedure.

GCA:　　Nippon Air 810, fly heading 330, descend and maintain 1,500.

PIL:　　Heading 330, descend and maintain 1,500, Nippon Air 810.

GCA:　　Nippon Air 810, contact final controller 119.5.

PIL:　　Contact Naha final controller 119.5, Nippon Air 810.

最終進入（ファイナルコントローラーとの交信）：着陸誘導開始前～着陸

１．通信設定

PIL: (Naha Final controller), Nippon Air 810.

GCA: Nippon Air 810, Naha final controller, how do you read.

PIL: Reading you loud and clear, (Nippon Air 810).

GCA: Do not acknowledge further transmissions.

２．グライドパス接近の通報（「approaching glidepath」）

GCA: Turn right heading 360, maintain 1,500, 6 miles from touchdown, approaching glidepath.

３．最終降下の指示（最終降下開始点に達したとき）

GCA: Begin descent, 5 miles from touchdown.

その後，必要に応じて，以下の４．～７．のような指示又は情報提供が行われる．

４．降下率の修正

GCA: Well below glidepath, adjust rate of descent, on course, 4 miles from touchdown. (*2)

５．最終進入コース（グライドパス）との関係位置・動き

GCA: Slightly left of course, turn right heading 002, coming back to glidepath slowly.

６．送信中断（トランスミッションブレイク）

　　　　最終進入中に適宜，管制官からの送信が５秒未満で中断される．必要があれば，この間にパイロットから送信を行ってもよい．(*3)

７．オン グライドパス

GCA: On glidepath, resume normal rate of descent.

８．着陸許可

GCA: Runway 36, cleared to land, wind 360 at 5 knots, correcting course nicely, now, on course, on glidepath, after landing contact Tower 118.1.

9．誘導の終了（*4）

GCA:　(Now) guidance limit, take over visually, if runway not in sight, execute missed approach.

１０．誘導終了後の位置情報（*5）

GCA:　Over approach light (threshold), on course, slightly below glidepath.

＊（*1）PAR Approach では，DA に達した点を誘導限界点とする．通常，滑走路進入端標高に 200 ft を加えた高度である．
　なお，周回進入の場合は，飛行場視認の通報が必要とされる．

GCA:　This will be a PAR approach to runway 36 for circling to runway 18, circling minimum altitude 500 feet, guidance limit ~ miles from touchdown, report airport in sight.

＊（*2）最終進入中の航空機に対しては，接地点からの距離を 1 マイルにつき 1 回以上通報される．

＊（*3）5 秒未満の送信中断は，故障等の通報・編隊で進入する航空機相互間の連絡等のために行われる．

GCA:　My transmission on final will be discontinued for less than 5 seconds, you may transmit during the period.
　　　（最終進入中の送信を 5 秒未満中断するから必要があれば送信して下さい）

＊（*4）次の場合に PAR Approach の誘導は終了する．
　　　・誘導限界に到達したとき
　　　・航空機からの要求（接地点からの距離を通報し，目視により進入するよう指示がある「～ miles from touchdown, take over visually」）

＊（*5）雲高の値が周回進入に係る最低降下高の値未満，又は，視程の値が周回進入に係る最低気象条件の地上視程の値未満に行われる．

　　　ヘディング及びグライドパスとの関係位置等に関する用語

on course	オンコース
slightly left of course	コースの少し左
well left of course	コースのかなり左
heading is good	針路良好
~ feet left of course	~ ft コースの左
on glidepath	オン　グライドパス
slightly above (below) glidepath	グライドパスより少し高い（低い）
well above (below) glidepath	グライドパスよりかなり高い（低い）
rate of descent is good	降下率良好
~ feet high (low)	~ ft 高すぎます（低すぎます）

（参考１）PAR Approach を行う前後の交信の一例を示す．

PIL:　　Tokyo Control, Nippon Air 810, FL 240.
ACC:　　Nippon Air 810, Tokyo Control, roger.

ACC:　　Nippon Air 810, descend to reach FL 200 by YARII.
PIL:　　Descend to reach FL 200 by YARII, Nippon Air 810.

PIL:　　Tokyo Control, Nippon Air 810, leaving FL 240 for FL 200.
ACC:　　Nippon Air 810, roger.

ACC:　　Nippon Air 810, contact Komatsu Approach 121.25.
PIL:　　Contact Komatsu Approach 121.25, Nippon Air 810.

PIL:　　Komatsu Approach, Nippon Air 810, FL 200.
RDR:　　Nippon Air 810, Komatsu Radar, runway 06, wind 030 at 5, QNH 3007, expect PAR approach.
PIL:　　Runway 06, 3007, expect PAR approach, Nippon Air 810.

RDR:　　Nippon Air 810, fly heading 250 vector to final approach course, maintain FL 200.
PIL:　　Heading 250, maintain FL 200, Nippon Air 810.

RDR:　　Nippon Air 810, descend and maintain 13,000.
PIL:　　Descend and maintain 13,000, Nippon Air 810.

RDR:　　Nippon Air 810, turn right heading 270.
PIL:　　Right heading 270, Nippon Air 810.

RDR:　　Nippon Air 810, descend and maintain 7,000.
PIL:　　Descend and maintain 7,000, Nippon Air 810.

RDR:　　Nippon Air 810, contact Komatsu Radar 134.1.
PIL:　　Contact Komatsu Radar 134.1, Nippon Air 810.

PIL:　　Komatsu Radar, Nippon Air 810, leaving FL 141 descending 7,000.
RDR:　　Nippon Air 810, Komatsu Radar, roger.

RDR:　　Nippon Air 810, descend and maintain 6,000.
PIL:　　Descend and maintain 6,000, Nippon Air 810.

RDR:　　Nippon Air 810, turn right heading 280, descend and maintain 5,000.
PIL:　　Right heading 280, descend 5,000, Nippon Air 810.

RDR:　　Nippon Air 810, turn right heading 300.
PIL:　　Right heading 300, Nippon Air 810.

RDR:　　Nippon Air 810, descend and maintain 3,500.
PIL:　　Descend and maintain 3,500, Nippon Air 810.

RDR:　　Nippon Air 810, turn right heading 340.
PIL:　　Right heading 340, Nippon Air 810.

RDR:	Nippon Air 810, this will be a PAR approach to runway 06, guidance limit 300 feet.
PIL:	Guidance limit 300 feet, Nippon Air 810.
RDR:	Nippon Air 810, turn right heading 360.
PIL:	Right heading 360, Nippon Air 810.
RDR:	Nippon Air 810, turn right heading 030.
PIL:	Right heading 030, Nippon Air 810.
RDR:	Nippon Air 810, turn right heading 040, descend and maintain 2,000.
PIL:	Right heading 040, descend and maintain 2,000, Nippon Air 810.
RDR:	Nippon Air 810, 18 miles southwest of Komatsu airport, perform landing check, wind 030 at 8.
PIL:	Nippon Air 810.
RDR:	Nippon Air 810, turn right heading 050.
PIL:	Right heading 050, Nippon Air 810.
RDR:	Nippon Air 810, Komatsu Final Controller, how do you read.
PIL:	Reading you five, Nippon Air 810.
RDR:	Nippon Air 810, do not acknowledge further transmission, turn right heading 055, right of course.
RDR:	Nippon Air 810, 12 miles from touchdown, turn right heading 060.
RDR:	Wind 040 at 7.

中略（P.157 のような指示が行われる）

RDR:	On course, guidance limit, take over visually, if runway not in sight, execute missed approach.
RDR:	Wind 040 at 7.
RDR:	On course, slightly above.
RDR:	Over threshold.
RDR:	Nippon Air 810, contact Komatsu Tower 118.25.
PIL:	Contact Komatsu Tower 118.25, Nippon Air 810.
PIL:	Komatsu Tower, Nippon Air 810, spot 23.
TWR:	Nippon Air 810, Komatsu Tower, turn left end of runway, contact Ground 121.7.

Phraseology Example 5

Surveillance Approach においては，レーダー誘導により高度を下げながら，最終進入コースに会合するよう誘導される．最終進入コースではヘディングの修正を指示され，降下パスについては滑走路進入端からの距離と適正高度の助言が与えられる．

最終進入開始前

1．誘導形式・滑走路・誘導限界が通報される

CTL:　This will be a surveillance approach to runway 36, guidance limit 1 mile from runway.（*1）

2．飛行場の気象情報（ATIS に含まれる場合は省略）

3．通信連絡途絶時の飛行方法の指示

CTL:　If no transmissions are received for 1 minute in the pattern or 15 seconds on final approach, attempt contact Tower 118.1 and proceed with VOR runway 36 approach.

4．進入復行方式（省略される場合もある）

CTL:　Your missed approach procedure is the same as the VOR runway 36 missed approach procedure.

最終進入

1．最低降下高度，降下中の適正高度の助言に関する通報

CTL:　Turn left heading 350, maintain 1,500, 7 miles from runway, surveillance minimum altitude 500 feet, guidance limit 1 mile from runway, recommended altitude will be furnished each mile on final until reaching the surveillance minimum altitude.

2．最終降下開始の予告と脚下げの注意喚起

CTL:　Turn right heading 360, prepare to descend in 1 mile, gear should be down.（*2）

3．最終降下の指示

CTL:　Coming back to course, turn left heading 358, 5 miles from runway, begin descent.

4．必要に応じコース修正・適正高度の助言

CTL: Well left of course, turn right heading 005, 4 miles from runway, altitude should be 1,200 feet.

5．着陸許可・最終進入コースとの関係位置及び動き

CTL: Runway 36, cleared to land, wind 040 at 6, you are still left of course and correcting slowly.

6．最低降下高度到達地点の通報

CTL: 1.5 miles from runway, altitude should be 500 feet, surveillance minimum altitude.

7．誘導限界高度に到達した旨の通報

CTL: Guidance limit, 1 mile from runway, take over visually, if runway not in sight, execute missed approach.（*3）

＊（*1）Surveillance Approach では，通常，滑走路進入端の1マイル手前の点を誘導限界点とする．誘導限界点を通報されるまでに滑走路が視認できなければ復行を行う．
＊（*2）「gear should be down」は指示ではなく，注意喚起の用語である（「perform landing check」も同様である）．
＊（*3）次の場合に Surveillance Approach の誘導は終了する．
　　　　・誘導限界に到達したとき
　　　　・誘導限界までの安全な進入が期待できないとき
　　　　・航空機からの要求又は滑走路の視認（滑走路端からの距離を通報し，目視により進入するよう指示がある「～ miles from runway, take over visually」）

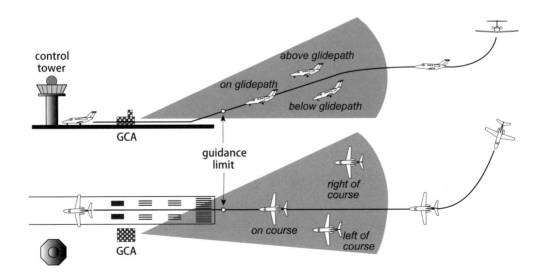

Phraseology Example 6

経路指定視認進入（CVA: Charted Visual Approach）の場合は，以下のようになる．経路指定視認進入とは，騒音軽減等を目的に，飛行すべき経路・高度及び顕著な地表物標（landmark）が図示された視認進入である（運航上の理由により，当該経路・高度を外れて飛行するのを妨げるものではない）．

PIL: Descend at pilot's discretion maintain 4,000, Nippon Air 154.

APP: Nippon Air 154, recleared via CREME direct CACAO, report highway in sight.

PIL: After CREME, direct CACAO, report highway in sight, Nippon Air 154.

PIL: Tokyo Approach, Nippon Air 154, highway in sight.

APP: Nippon Air 154, cleared highway visual runway 34R approach.

PIL: Cleared highway visual runway 34R approach, Nippon Air 154.

APP: Nippon Air 154, contact Tower 124.35.

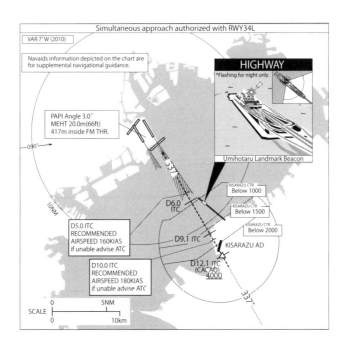

　CVA では，パイロットが物標（landmark）を視認したことを通報し，先行進入機の視認状況により CVA が許可される．また，関連航空機を視認したか否かでタワーへの通信移管のタイミングが変わる．

１．先行機がない場合

　物標（landmark）を視認したことを通報した後に CVA が許可される．また，この時点以降，適切なタイミングでタワー に移管される．

２．先行機がある場合

　２－１．先行進入機（同一滑走路への進入機）を視認した場合

　　　Landmark を視認し，かつ先行進入機を視認した旨通報した場合は，目視間隔を維持するよう指示され CVA が許可される．また，この時点以降，適切なタイミングでタワーに移管される．

PIL:　　Tokyo Approach, Nippon Air 154, highway in sight.

APP:　　Nippon Air 154, cleared highway visual runway 34R approach.

PIL:　　Cleared highway visual runway 34R approach, Nippon Air 154.

APP:　　Nippon Air 154, preceding traffic Boeing 737 7 miles ahead, report traffic in sight.

PIL:　　Report traffic in sight, Nippon Air 154.

PIL:　　Tokyo Approach, Nippon Air 154, traffic in sight.

APP:　　Nippon Air 154, maintain visual separation from traffic, contact Tower 124.35.

　　２－２．先行進入機が視認できない場合

　　　Landmark を視認し，先行進入機を視認できない旨通報した場合は，先行進入機の位置情報を通報後 CVA が許可される．この場合，タワーへの移管は，自機が先行進入機を視認した旨通報した時点若しくは，タワーが自機と先行進入機をともに視認し目視間隔を設定できると判断し，アプローチにその旨通報した時点で行われる．

PIL:　　Negative contact, Nippon Air 154.

APP:　　Nippon Air 154, preceding traffic 8 miles ahead, cleared highway visual runway 34R approach, report traffic in sight.

先行進入機が視認できた場合

PIL:　　Tokyo Approach, Nippon Air 154, traffic in sight.

APP:　　Nippon Air 154, maintain visual separation from Boeing 737, contact Tower 124.35.

タワー が自機と先行進入機を視認し目視間隔を設定できると判断した場合

APP:　　Nippon Air 154, Tower has you in sight, contact Tower 124.35.

捜索救難（RCC）発動基準

緊急状態	緊急状態を知った管制機関
不確実の段階 (Uncertainty Phase: INCERFA)	
1　位置通報又は運航状態通報が予定時刻から 30 分過ぎても ない場合 2　航空機がその予定時刻から 30 分（ジェット機にあっては 15 分）過ぎても目的地に到着しない場合	1　第 1 段通信捜索を行う. （注 1） 2　救難調整本部（RCC）に 通報する. 3　可能ならば当該航空機の 使用者に通報する.
警戒の段階 (Alert Phase: ALERFA)	
1　第 1 段通信捜索で当該航空機の情報が明らかでない場合 2　第 1 段通信捜索開始後 30 分を経ても当該航空機の情報が 明らかでない場合 3　航空機が着陸許可を受けた後，予定時刻から 5 分以内に着 陸せず当該航空機と連絡がとれなかった場合 4　航空機の航行性能が悪化したが，不時着のおそれがある程 でない旨の連絡があった場合	1　拡大通信捜索を行う. （注 2） 2　捜索救難に必要と認めら れる情報又は資料を RCC に通報する. 3　可能ならば当該航空機の 使用者に通報する.
遭難の段階 (Distress Phase: DETRESFA)	
1　拡大通信捜索で当該航空機の情報が明らかでない場合 2　拡大通信捜索開始後 1 時間を経ても当該航空機の情報が明 らかでない場合 3　当該航空機の搭載燃料が枯渇したか，又は安全に到着する には不十分であると認められる場合 4　当該航空機の航行性能が不時着のおそれがある程悪化した ことを示す情報を受けた場合 5　当該航空機が，不時着をしようとしているか，又は既に不 時着を行った情報を受けたか若しくはそのことが確実であ る場合	収集した情報を RCC に通報す る.

注 1　第 1 段通信捜索とは，計器飛行方式による航空機については，その予定経路上における同 機と交信し得る管制機関の有する施設を利用して行う捜索をいい，有視界飛行方式による 航空機については，その予定経路上における飛行場について行う捜索をいう.

注 2　拡大通信捜索とは当該航空機の到着可能な範囲にある関係機関による捜索をいう.

交信上の主な語彙・表現（右側は該当ページ）

POINT 一覧（右側は該当ページ）

Memo

Memo

Memo

■━━━━━━━━━━［著者］━━━━━━━■

縄田義直(Yoshinao Nawata)

1976年熊本県生まれ.

一橋大学大学院言語社会研究科博士後期課程退学.

現在, 独立行政法人航空大学校教授. 専攻は社会言語学・航空英語.

著書に『航空留学のためのATC』/共著,『ATC入門－VFR編－』,

『ATC入門 －リスニング編－』(ともに鳳文書林出版販売発行)など.

平成24年1月30日　初版発行　　　　　　　　　　　　　　　印刷　シナノ印刷㈱
令和5年4月18日　第4版発行

ATC入門 －IFR編－

縄田義直著

鳳文書林出版販売㈱

〒105-0004　東京都港区新橋 3－7－3

Tel 03-3591-0909　　Fax03-3591-0709　E-mail info@hobun.co.jp

ISBN978-4-89279-474-2　C3032　Y 3300E　　　　　定価 3,630円（本体価格 3,300円＋税10%）